"Once you are convinced that the only way we are ever going to be able to study and understand life as it is actually lived is through ethnography, then you know why it is so important for other people to appreciate this and ideally learn to be ethnographers. Reading this book you can feel the passionate commitment to this goal of persuasion. Thanks to its abundant 'story box' illustrations, its discussion of foundational principles and practices and above all the sense of a holistic imagination of what this holistic method entails, the book effectively delivers. Ethnography will grant you empathy and expand your life, but first, you need to learn what it is."

Professor Daniel Miller, *Professor of Anthropology, University College London*

ETHNOGRAPHY
THE BASICS

Ethnography: The Basics introduces a broad and beginner audience to ethnography as a research methodology with diverse applications. By using everyday language and developing a warm and inclusive tone, the book provides an accessible entry point to the topic.

It offers a picture of the practice of ethnography that is both human and humane, tackling some of the practical barriers, ethical complexities, and lived experiences in an honest and straightforward way. Organised into eight chapters, the book provides a comprehensive overview of the unique purpose, value, and scope of ethnography, as well as a practical guide as to what the practice of ethnographic research involves, such as steps for designing research, gathering data, and analysing and presenting findings.

Without relying on assumed familiarity with or interest in discipline-specific histories and frameworks, *Ethnography: The Basics* combines clear structure, plain language, and thoughtfully selected examples and stories to welcome students into this interesting and valuable subject. The book will also be of great interest to academics and professionals who wish to supplement their ethnographic knowledge.

Susan Wardell is an academic and writer from Aotearoa New Zealand. She is Senior Lecturer in the Social Anthropology programme at the University of Otago, with a shared background in communication studies. Her own research focuses on emotion and affect, digital sociality, health, and mental health, with recent work covering topics such as online medical crowdfunding and ecological distress. She is interested in public communication and in creative ethnography, writing and publishing in several literary genres, while also dabbling in visual and performance arts. Susan is Pākehā (New Zealand European) and lives in Ōtepoti Dunedin.

THE BASICS

The Basics is a highly successful series of accessible guidebooks which provide an overview of the fundamental principles of a subject area in a jargon-free and undaunting format.

Intended for students approaching a subject for the first time, the books both introduce the essentials of a subject and provide an ideal springboard for further study. With over 50 titles spanning subjects from artificial intelligence (AI) to women's studies, *The Basics* are an ideal starting point for students seeking to understand a subject area.

Each text comes with recommendations for further study and gradually introduces the complexities and nuances within a subject.

For more information about this series, please visit: www.routledge.com/The-Basics/book-series/B

ETHNOGRAPHY

THE BASICS

Susan Wardell

Routledge
Taylor & Francis Group

LONDON AND NEW YORK

Designed cover image: Photograph by the author, Susan Wardell

First published 2025
by Routledge
4 Park Square, Milton Park, Abingdon, Oxon OX14 4RN

and by Routledge
605 Third Avenue, New York, NY 10158

Routledge is an imprint of the Taylor & Francis Group, an informa business

British Library Cataloguing-in-Publication Data
A catalogue record for this book is available from the British Library

ISBN: 9781032513119 (hbk)
ISBN: 9781032520124 (pbk)
ISBN: 9781003404880 (ebk)

DOI: 10.4324/9781003404880

Typeset in Times New Roman
by Apex CoVantage, LLC

For my students, with whom I have learned just as much about ethnography as I have through my own research practice.

CONTENTS

FIGURES

BOXES

ACKNOWLEDGEMENTS

I would like to thank and acknowledge my mentor, Professor Ruth Fitzgerald, who first introduced me to ethnography and then supervised (so wisely, kindly, and enthusiastically) my own earliest efforts.

Thank you to the reviewers of the proposal and first draft for this book, who made incredibly valuable comments, and especially to Lorena Gibson. Thank you to Anna Williams for feedback on some early chapters, and to Terin Hastelow for encouraging comments on my opening chapter, which I found the hardest of all to write. Thank you to my mother, Marilyn Smirk, for unwavering support and for help with all of those unruly commas.

Thank you most heartily to my other family members, friends, and colleagues near and far for bearing with my distraction and occasional frustration as I worked to complete the book amidst the ongoing hustle of work and life.

GLOSSARY

Following are some simple working definitions of key terms related to ethnography and that are used frequently in this book. These are bolded the first time they appear in each chapter.

Autoethnography/autoethnographic: A type of ethnography in which the ethnographer makes themselves and their own personal experiences the subject of analysis by connecting these to wider (social, cultural, political, economic) contexts

Bracketing: A process of temporarily setting aside some aspect/s of your pre-existing knowledge, beliefs, or expectations

Coding: A process of attaching descriptive labels (or codes) to different bits of data in order to look for patterns

Collaborative: Two or more individuals or groups working actively and intentionally together towards a particular goal, process, or output

Comparative: Building understandings by observing the similarities and differences between two or more things

Critical: A lens focused on questioning why things are the way they are, often through deconstruction, and with a focus on understanding how they relate to wider institutions, systems, and power structures

Cultural relativism: A way of making sense of the cultural practices and norms of others within their own context, rather than comparing them to the norms of your own context (opposite: **ethnocentrism**)

Culture: A system of shared meaning that may include visible features (such as traditions, forms of dress, dialect or language, holidays or events) but also invisible ones (such as norms, values, identities, systems of social organisation)

Decolonising/decolonisation: Practices or processes that are about deliberately undoing colonial and colonised ways of thinking, acting, or organising the world

Deductive: Having a pre-existing idea or hypothesis that you test via gathering data (opposite: **inductive**)

Defamiliarisation: Focusing on something that is usually taken for granted or familiar and allowing yourself to see it in a new or different way

Emic: An insider perspective (opposite: **etic**)

Empirical: Data based on observation or direct experience through the senses

Epistemology/Epistemological: The study of (or philosophy of) knowledge, including its sources, scope, methods, and measures of validity

Ethnocentrism: A way of evaluating other people's practices and norms through the lens of your own cultural frameworks (opposite: **cultural relativism**)

Ethnographic lens: A distinctive way of approaching social and cultural phenomena, focused on descriptive, situated, holistic knowledge and on reflexive and collaborative processes

Ethnographic record: The cumulative knowledge recorded and shared from prior ethnographic research

Etic : An outsider perspective, specifically based in an analytic or academic framework (opposite: **emic**)

The Field: The social space, community, or setting that an ethnographer has chosen as the focus of their research

Fieldnotes : A type of text that ethnographers produce themselves from observations and experiences in the field

Fieldsite: A specific place or setting in which the ethnographer is engaging with participants and gathering data

Fieldwork: Any form of data gathering focused on gaining insights from time spent with people in real-world settings

Gatekeeper: A person who mediates or controls the access a researcher has to the field or to other participants

Generalising/Generalisation: Applying an idea, principle, or theory broadly or beyond the context in which it was initially identified

Holism/Holistic: A lens emphasising how different parts of the social world are interconnected and therefore cannot be understood without reference to the whole

Humanistic: A framework or lens that aims to recognise and honour human values and agencies, as central to the relationships and meaning-making practices that make up the social world

Inductive: Allowing meanings, understandings, or conclusions to be derived from data as it is gathered and analysed (opposite: **deductive**)

Informant: A person who provides information about the research topic, based on being embedded in the field themselves. Also called a participant, or interlocuter.

Insider ethnography: Ethnographic research in which the researcher is already a member of the community or group being researched, prior to commencing research

Interlocuter : A research participant or subject; someone you talk to as part of gathering data

Interpretive/Interpretivist: Understanding knowledge as actively constructed through the researcher's own sense-making processes, not simply 'discovered' (opposite: **positivist**)

Iterative: Moving back and forward through different stages or steps of the research process multiple times or in a nonlinear fashion

Key informant: A person who turns out to be particularly significant as a source of knowledge and insight in a study

Liminality: A state or quality of being 'betwixt and between'

Method/Research method: A specific technique or process for gathering data

Methodology/Research methodology: A set of principles, values, or philosophies, which explain your particular approach to research, including what methods you use

Multimethodological: Research that uses more than one method to gather data

Multimodal: A project or text that draws on multiple different senses, modes, or mediums of communication at the same time

Multisited: Using more than one fieldsite for data collection

Naturalistic: Something examined in an everyday or 'real-world' setting, as opposed to a controlled or artificial setting

Objective/Objectivity: An approach to establishing knowledge which aims to remove any personal, social, or cultural relationship to the subject matter, based on the premise that facts can exist independently to the person observing them (opposite: **subjective**)

Ontological: To do with the nature of being and existence

Othering: Emphasising the difference or 'otherness' of a person or group in a negative way, or a way that creates social distance or exclusion

Participant: A person who agrees to be part of a study and engage with the researcher; the consenting subject of a research project

Participant-observation: A technique for gathering data that involves taking part in a social world, setting, or practice in order to record what you observe and experience

Positionality/Social Positionality: How someone is located within the wider social, political, cultural world, usually including factors such as ethnicity, nationality, gender, race, socioeconomic class, dis/ability, geographic location, citizenship status, education level, and so on

Positivist: A lens or framework in which reality is seen as having fixed qualities, independent of human observation, and research is understood to be about scientifically determining these realities

Postmodern: A philosophical or intellectual stance reacting against the ideas of modernism and emphasising deconstruction, reflexivity, intertextuality, questioning of truth and authority, and subjectivity

Qualitative: Related to describing something's qualities and characteristics (opposite: **quantitative**)

Quantitative: Related to measuring something (opposite: **qualitative**)

Reductive/Reductivist: Presenting something complex or nuanced in an oversimplified way

Reflexive/Reflexivity: A process of deliberately and critically examining your own positionality and practices in terms of how these might shape your engagements in knowledge-making

Sampling: The logic behind who or what you choose to include in your dataset

Situated: Having qualities specific to the context it is embedded in

Subject/Research subject: A person who is the focus of research inquiry. Also called a participant, informant, or interlocutor

Subjectivity: A way of being in the world (and/or a stance from which knowledge is made) that is specific to a historical, cultural, economic, political and embodied positionality (opposite: **objective)**

Temporality: The experience of being in or relating to the movement of time

Theory: A set of interconnected concepts or ideas that describe something about the how the world works

Thick description: Descriptions that focus on identifying the context, meaning, and significance of what is being described, not just providing the literal observable facts

Triangulation/Data triangulation: Connecting across or between multiple different sources of data

Vignette: A brief, evocative account of some sort of observed happening

Worldview: A framework (of attitudes, values, stories, beliefs, and expectations) that shapes how someone understands reality

The index at the end of the book (see pg 239) also offers a way to trace more specific topics as they appear throughout the book, based on what pages they appear on.

WHY ETHNOGRAPHY?

We are all participants in the social world: a complex, vibrant place in which we live our lives, interact with others, and find and make meaning. The social world is also not *a* world but *worlds*, diverse, plural, and overlapping. Ethnographic research can be used to both describe and analyse this. It does so at the level of communities, **cultures**, organisations, institutions, subcultures, social movements, and societies, providing a way to examine them not only in terms of what can be observed from the outside but with a focus on the qualities of 'life as lived'. To do this, ethnographers have techniques that allow them to immerse and to participate as well as to observe, record, and ask questions. This creates opportunities for very human sorts of encounters where researchers can learn from people in the context of their everyday lives and where understandings can be built up cumulatively and **collaboratively** over time.

Ethnographers pay attention. They listen deeply. They question the taken-for-granted. They tell stories.

This book aims to help the reader further their understanding of the purpose and scope of ethnography (*what ethnographers are interested in, what they respond to, and why their work matters*) as well as providing a practical guide to what the actual practice of ethnographic research involves (*steps for designing research, gathering data, analysing findings, and constructing texts*). In this shorter introductory chapter, I start things off by discussing what defines ethnography as a research methodology. I provide some background on the distinctive features and methods of ethnography, the way it has changed over time, some of the practical pros and cons, and the bigger questions of what makes it useful in a contemporary world.

DOI: 10.4324/9781003404880-1

I conclude the chapter with an introduction to myself (Susan, the author), an outline of the following chapters, and some ideas about how to make the most out of the book.

> The term 'ethnography' refers to both a type of research and a genre of text.

WHAT (AND HOW) DO ETHNOGRAPHERS RESEARCH?

Ethnography is part of a **qualitative** and **interpretive** research tradition. It has its roots in the discipline of social/cultural anthropology but today has been taken up in a variety of different fields. It is used not just in academic or scholarly spheres but also in the public sector, the commercial sector, for industrial design, as part of community-based or activist projects, and more.

> People who use an ethnographic methodology to conduct research are called 'ethnographers'.

An ethnographic approach allows researchers to zoom in to the details and textures of a particular social setting in order to learn more about the people in it: what they say, what they do, how they interact, how they feel, and what matters to them. At the same time, ethnographers are interested in understanding the social practices and experiences they encounter as part of wider social, historical, and political contexts. This means they also zoom out to analyse, deconstruct, and theorise how different parts of the social world are related and connected, in what ways they might vary across different settings, and why they either change or persist over time. These interests can be applied to a vast array of topics, including work, art, conflict, health, morality, technology, family, education, activism, business, spirituality, violence, . . . and anything else human. Robert Desjarlais's ethnographic research, for example, focused on shamanic healing in the Nepalese mountains (see pg 97). Ned Barker & Carey Jewitt used ethnography to examine the impact of robots on industrial workers in the UK (see pg 83). Albert Refiti wove ethnographic methods into his study of the Pasifika concept of 'Vā' in architecture and design (see pg 113). Sahana Udupa conducted an ethnographic study of news media organisations in the global city of Bangalore (see pg 23). These studies may seem wildly different to one another, but what is shared by all of them is a particular *methodological* approach.

THE ORIGINS OF ETHNOGRAPHY: METHODOLOGY AND METHOD

Ethnographers are interested in social and cultural worlds: in norms, values, identities, relationships, practices, power, meaning, and social organisation. But the scope of this is vast, and ethnography is defined less by *what* it studies and more by *how* it studies. In the early 1900s, in an era of colonial science, ethnography emerged in European and North American academic institutions as a radical new approach to studying human cultures. It provided a more systematic and scientific approach to this than what had come before. But ethnography was novel in another way too: in suggesting that scholars go out and gather data themselves through **fieldwork** (see also Chapter 2). This approach was grounded in the idea of **naturalistic** research, which focuses on research conducted in real-world settings. Going even further, ethnography proposed that researchers not just make brief observations but rather spend significant periods of time with the people they wanted to learn about, immersing themselves in their lives. This stretched the existing idea of scientific 'observation' so much that it generated a new term for what ethnographers did: **participant-observation**. The goal was to complement what an objective observer's perspective might offer with efforts to understand an insider's point of view (Robben and Sluka 2012).

These innovations, from inside the academy, opened a pathway to valuing more diverse forms and sources of knowledge. Specifically, it opened the door for ethnographers to prioritise experiential and firsthand knowledge from 'being there'. Even more importantly, it opened the door for researchers to work collaboratively with **participants** they had the time to build relationships with, recognising and valuing the knowledge people already have about their own lives. Ethnographers often did this by working with local informants, research assistants, or translators, including with families they stayed with or with friends they made in **the field**. The more detailed and nuanced accounts contributed to the ability of social scientists to do **comparative** work, but in a specific way: through the lens of **cultural relativism**, which asked researchers and their audiences to de-centre their own familiar or taken-for-granted understandings and instead work to interpret what they were seeing in its own context.

While fieldwork and participant-observation remain distinctive markers of ethnographic research, contemporary ethnographic projects can also be diverse among themselves, drawing on a flexible toolkit of techniques for gathering and analysing data, and representing research findings. For example, Susan Jane Lewis and A.J. Russell embedded themselves in a public organisation devoted to tobacco control in England, where they shadowed the organisers' activities, observed how they networked with other stakeholders, and studied the various documents they produced (see pg 56). Ella Fisher and Alex Nading organised a series of workshops with people from a particular urban community in Nicaragua, where they used playful and interactive activities to speculate about what a thriving future would look like (see pg 164). Greta Blackwell met with successful professionals, in their homes, to facilitate group conversations about a popular reality TV show, and why it mattered to them (see pg 161). Andrew Gilbert and Larisa Kurtović collaborated with workers at a Bosnian detergent factory and with a graphic artist to produce illustrations about a landmark instance of union activism (see pg 209). Tarapuhi Vaeau used an **autoethnographic** and Indigenous 'kaupapa' approach to build her understandings of Māori responses to historical trauma through time spent with her own extended family (see pg 71). What brings all of these different projects together, despite their different topics *and* diverse techniques, is something best described as an **ethnographic lens**.

A research **method** is a particular technique or process for gathering and/or analysing data. A research **methodology** is a system of principles, values, and philosophies that explain why you are doing research a certain way, including the logic behind what methods you are using.

THE ETHNOGRAPHIC LENS

Ethnography is defined less by a strict set of procedures and more by a way of looking at and making sense of the world. Because of this, although it does have some favoured or distinctive methods, ethnography is overall better viewed as a *methodology*: a framework of principles which inform every part of the process, from designing projects to gathering and analysing data to constructing texts. Even more than this, it can be seen as a *lens* – whose core features can be distilled into the following:

Descriptive: Ethnographic research places value on providing detailed qualitative descriptions of a social field. This is achieved through a focus on recording the everyday minutia of life, and through techniques that aim to convey the vividness and texture of life as lived as well as observed.

Situated: Ethnography focuses on building in-depth and contextualised knowledge of *particular* people in *particular* places, times, and settings rather than **generalising** on a large scale. The focus of ethnographic projects can appear quite small or 'local', but this allows ethnographers to explore how more global or universal phenomena may be articulated differently in different settings.

Holistic: Ethnographers recognise that the different parts of the social world are all interconnected – for example, how systems of kinship might be connected to economics, politics, and law or how aesthetic design might be related to gender, ideology, and work. They emphasise that the details of whatever is observed must be thought about as a part of a larger social whole, showing also then how personal or everyday things are connected to wider social systems and structures.

Collaborative: Ethnography focuses on producing knowledge *with* instead of just *about* people. It relies on participants who are willing to open up their lives or share their own understandings in some form. It therefore asks researchers to consider how best (within the parameters of their own **fieldsites**) to connect, to listen, and to honour the voices, knowledge, and agency of their participants.

Reflexive: Ethnographers are trained to reflect critically upon how knowledge is produced. This means being aware of and transparent about the specific actions and encounters that their findings are based on, and how their own social **positionality** might shape their ways of gathering and interpreting data. These habits of reflexivity help them to navigate sensitive and ethical engagements in complex social settings, and to recognise how power is involved in their own practice and in broader knowledge institutions.

Something is described as 'ethnographic' if it contains the hallmarks of this unique way of seeing, understanding, and analysing the social world.

TURN, TURN, TURN: CRITICAL TRANSFORMATIONS IN ETHNOGRAPHIC PRACTICE

Ethnography has always held within it some radical potential, but its origins in the Western academy, and in colonial ways of thinking and researching, left plenty to critique. This critique was not just about the practices of individual researchers, but about some of the **ethnocentric** assumptions of the methodology as a whole. Change started with a shift in the focus of *who* and *what* and *where* ethnographers were researching (see Chapter 2). But this occurred alongside debates about the **epistemological** basis of ethnographic knowledge.

In early eras, scholars fought to gain legitimacy for ethnography as an academic method through aligning it with **positivist** scientific paradigms. This focused on ethnographers making authoritative knowledge claims about the people they were studying. A shift in thinking called the interpretive turn, from around the 1970s onwards, helped with reforming some of this. While remaining focused on gathering **empirical** data, ethnography began to turn more towards an interpretivist paradigm. The ethnographer's job came to be seen as being less about recording **objective** facts and more about interpreting webs of meaning. As well as setting up an important tradition of reflexivity, this established a comfort with the open-endedness of many questions about the social world and a willingness to leave room for complexity, **subjectivity**, ambiguity, and the "fuzzy, unscientific qualities" of life itself (Wolcott 2005). As such, ethnography came to be described as "the most **humanistic** of the sciences, and the most scientific of the humanities" (Van Maanen 2006, p. 242).

Ethnographers continued to deconstruct their practices through a series of movements, crises, and 'turns'. Major things were at stake throughout all these conversations, including the professional identity of ethnographers, and the legitimacy of ethnographic knowledge. At the same time, these debates generated new innovations. The political and social movements in the 1960s and 1970s – an exciting time of reckoning with power relations in research and in the wider world and for questions about how ethnographers might align their work with social causes – led to the development of **critical** ethnography and feminist ethnography and to work that could be described as activist ethnography. This also fanned the flames of ongoing work towards **decolonising** ethnography. In the 1980s and 1990s, the **postmodern** turn and the **reflexive** turn, while pulling apart some of the traditional bases of ethnographic authority and questioning many of the representational norms of ethnographic texts, supported the continuing development of autoethnography and a variety of forms of experimental ethnography.

Ethnography has come a long way through its practitioners taking up difficult conversations. Even so, it is not that ethnography has been 'refined to perfection'. Ethnographers continue to grapple with some of the previously mentioned questions in an imminent way, even as fresh challenges emerge. Meanwhile, further 'turns' in the social sciences and beyond have invigorated ethnographic thinking and practice, including (but not limited to) the linguistic turn, temporal turn, affective turn, **ontological** turn, literary turn, emotional turn, somatic turn, creative turn, and post-human turn. These many turns and tides have resulted in a commitment to deep, engaged, critical, and reflexive conversations and an openness to new learning opportunities.

PUTTING ETHNOGRAPHY TO WORK

Although it continues to have strong associations with social and cultural anthropology, ethnographic research methodologies have steadily been taken up in more and more academic, professional, and applied fields including sociology, psychology, geography, education, health research, tourism, management studies, nursing, sports studies, and many more. It has also become popular in consumer research, UX/user experience, and design.

PRACTICAL ADVANTAGES AND LIMITATIONS

Ethnographic research takes considerable skill and practice to do well, but other practical factors make it an accessible methodology for a variety of people to use. Firstly, though there are plenty of settings in which ethnographic research may be conducted by a research team, it can also be conducted by an individual researcher, and in fact this remains the most common approach. Secondly, ethnographic research doesn't typically require any expensive or specialised equipment, so it can be a low-cost form of research. On the other hand, ethnographic research tends to be incredibly time intensive, both in terms of projects that span long periods of time *and* practices based in intense or immersive engagement. Because ethnographic data gathering relies on a researcher's direct presence, observation, or involvement in the fieldsite, this work can't be delegated or distributed to others easily.

In addition, there are many factors that a successful project may depend on which are outside of the control of the individual researcher – including access to certain spaces or fields and the responsiveness or availability of the participants. This means ethnographic projects often rely on a degree of flexibility and can be challenging to conduct when there are strict parameters around timelines, outputs, or outcomes. All of this requires ethnographers to show resilience as well as humility, to think on their toes, and to be able to adapt. They must do this while recognising that they are also dealing with real people and their actual lives, meaning there can be no shortcuts to doing ethnographic research ethically and responsibly.

DIVERSIFICATION, ECLECTICISM, AND 'BORROWING ETHNOGRAPHIC METHODS'

The diversity of techniques that ethnographers may choose from can be an incredible strength, providing rich potential to make the work applicable for a variety of communities, audiences, or clients. It also means that there are potentially "infinite variations" (Wolcott 2005, p. 15). In addition, ethnography is often shaped by the discipline of the researcher (Pink and Morgan 2013, p. 352). The resulting eclecticism of practice has caused anxiety for some people: *How do we know what counts as ethnography? How will others know what we mean when we say it?* It is also true that because ethnography requires a lot of time investment, researchers in some fields may "borrow ethnographic techniques" (Wolcott 2008), meaning they adapt some of the tools or strategies of ethnography to suit their own constraints: for example, doing more short-term stints of participant-observation, gathering people for focus groups in a community context, or asking participants to share insights about certain parts of their lives in mediated forms. Not all of these practices would necessarily meet the criteria of a traditional ethnographic study, and some people argue they should be kept distinct from 'doing ethnography' in response to worries about the fuzzy boundaries of the practice and in order to keep building its legitimacy as a research methodology (Wolcott 2008, p. 44). However, other scholars have used this as an opportunity to further discuss what might connect these approaches.

Feminist scholar Donna Haraway has suggested an 'ethnographic attitude' as a "mode of practical and theoretical attention" which could be adopted in any type of enquiry as a way of remaining mindful of and accountable to the broader principles of ethnography (Haraway and Goodeve 2018, p. 191). Martin Gerard Forsey, who studies public education, writes about the 'ethnographic imaginary' as a commitment to the sort of deep listening and close observation that might assist with paying attention to the social world in a very particular way (2010, p. 567). Both of these ideas resonate with the idea of an ethnographic 'lens'.

WHY ETHNOGRAPHY MATTERS

Why does all of this matter? What is ethnography useful for?

Contemporary ethnographic research asks human questions about human problems. It contributes to documenting the incredibly diversity of human social life – based on a warm-hearted curiosity about the ways people live out what matters to them amidst many different pressures and within specific (political, organisational, geographic, technological, economic, and cultural) contexts. The type of knowledge ethnographers produce in collaboration with their participants is not simplified or short-form. It is nuanced and complex. But it also retains engaging qualities and the ability to speak to things of pressing importance to public life. As anthropologist Alpa Shah writes, ethnography can be a revolutionary praxis because it "makes us question our fundamental assumptions and pre-existing theories about the world; it enables us to discover new ways of thinking about, seeing, and acting in the world" (2017, p. 47). As such, ethnography has a variety of different purposes and applications, including:

- For people who want to contribute to scholarly theorisations of the social and cultural world in all its beauty, power, and strangeness
- For people who want to apply these sorts of knowledges at the level of organisations, communities, or governments to work for positive change
- For people who want to design products or services that respond well to the actual needs of actual people
- For people who want to tell vivid and meaningful human stories and engage wider audiences in them
- For people who want to explore the speculative possibilities of what the world *could* be, asking open-ended questions, and stirring up curiosity, reflection, and perhaps even hope

ABOUT THE BOOK AND HOW TO USE IT

This book is both informational and practical. It lays out some background around ethnography's key concepts, definitions, and debates *and* offers considerations for someone who is conducting their own ethnographic project. It is aimed at beginners. That said, ethnography is a multifaceted research methodology, and even the most experienced ethnographer is learning all the time, so even if you've done this sort of research before you will likely find new angles enlivened here. Keep in mind that the book can only offer a brief overview of most topics, so there will also be a recommended reading list at the end of each chapter to help you go deeper if you wish to.

The book is aimed at an interdisciplinary audience. The origins of ethnography are very much tied to the academic discipline of social/

cultural anthropology, so many books or resources about ethnography assume readers have training in that area. But ethnography is used widely now, so this book aims to introduce the subject in a way that doesn't rely on you having an anthropological background. This means I try to avoid assuming you know key concepts or terms already, and instead, keywords are bolded throughout each chapter, and the glossary (on p. xxiii) gives you definitions for these in plain language.

ABOUT THE AUTHOR

This might be a good moment to introduce myself – Susan, the author of this book. I am a full-time academic, involved in my own research, in teaching undergraduates and supervising postgraduates, and in writing, editing, and publishing. I live and work in Aotearoa, New Zealand. I am Pākehā (a New Zealander of European heritage), with ancestors from Australia, England, Ireland, Scotland, and Germany, who was privileged to grow up in the beautiful city of Ōtepoti Dunedin, as Tangata Tiriti (a settler; a person of the Treaty of Waitangi). My academic training is in social anthropology and communication studies, and I have a PhD from the University of Otago, where I am currently a senior lecturer in the Social Anthropology programme. I love the discipline I teach in and, at the same time, am very excited by interdisciplinary and multidisciplinary work. I am also a practitioner of creative writing in a number of forms – most often poetry, flash fiction, and essay – and an enthusiastic dabbler in both visual and performance arts.

I have had modest amounts of experience with a variety of types of ethnography, including multisited fieldwork, ethnography both 'at home' and overseas, organisational ethnography, comparative ethnography, digital ethnography, interview-based research, applied projects, creative ethnography, visual ethnography, and more. The work I have done using these sorts of methods represents only a very small sample of the vast possibilities for ethnographic research. In addition, they are grounded in my experiences as a white, cis female, English-speaking ethnographer based at a tertiary institution. No one person can speak to all the different dynamics of ethnography, and the choice not to go much into detail about my own work throughout this book, despite the important reflexive tradition of ethnography, is mainly so that I can spend the limited wordcount establishing a

Figure 1.1 The author, Susan Wardell, sitting on some mostly-demolished stairs, during a fieldwork trip to Ōtautahi Christchurch (Aotearoa New Zealand), February 2021.

broader snapshot of ethnographic experience and practice. I do this by illustrating (and celebrating) the diversity of other people's stories, research, and writing in 'story boxes' throughout.

THE STRUCTURE AND CONTENT OF THIS BOOK

The book is structured into nine chapters that build on each other in a strategic manner but can also be read individually. Chapter 2 starts by getting you thinking about the 'field', often a first decision in designing in ethnographic research project but also an idea which is slippery and entangled with changes in the history of ethnographic practice. Chapter 3 spends some time acknowledging that ethnographic research is relational at its core and exploring the many ethical, personal, and institutional complexities in this. Chapter 4 focuses on 'being there' as a participant-observer, charting some ways you might orient your mind, body, and heart to this task. Chapter 5 addresses ways of recording or producing data including fieldnotes, drawing, visual, sounded or audio-visual recording, and the use of existing records or texts. Chapter 6 follows this up with a focus on the art of listening and asking questions in conversations, interviews, and group settings, also briefly teasing out the possibilities for collaborative and arts-based techniques. Chapter 7 thinks about how we make sense of the data we end up with through processes of organising, coding, analysing, interpreting, and theorising. Chapter 8 recognises that no ethnography is finished until it takes some kind of textual form and wrestles with some of the epistemological and ethical complexities of representing ethnographic knowledge and being responsible for the impact it makes in the world. The book concludes with a brief Epilogue that is my own personal 'love letter to ethnography'.

You won't find any chapters or sections of the book specifically devoted to ethics or to the 'behind-the-scenes' emotional experience of the ethnographer for the important reason that these are threaded throughout *all* chapters. I have tried to signpost these clearly where they do come up, and if you are looking especially to read about ethics, Chapter 3 is the most direct 'in'. One of my main goals has been to grapple honestly with complexity of ethnography as a form of qualitative inquiry – the humanness and messiness of it all. Despite this (or perhaps because of it), for many of us, "It is ethnography that keeps us open to the world and provides the insights we return to the world" (Miller 2017, p. 30). Because of this I am hopeful that you will be left with a picture of what a unique, valuable, and fulfilling approach ethnography offers, as well as a sense of what possibilities it might hold for you specifically.

At the end of the day, ethnography is about people: about ourselves and those we share the world with. For those who are hungry to explore this – for those who want to observe the world, or change the world, or both – ethnography is for you, and I hope this book will whet your appetite. Shall we begin?

CHAPTER SUMMARY: KEY POINTS

- Ethnography is a qualitative and interpretive research tradition. It focuses on in-depth explorations of the social world at the level of communities, cultures, organisations, institutions, subcultures, social movements, and societies.

- Fieldwork and participant-observation are the distinctive methods of ethnographic research. In practice, ethnographers draw on a flexible range of techniques for recording, analysing, and communicating data.
- Ethnography is simultaneously a research methodology and a type of text, both defined by a 'lens' that emphasises descriptive, situated, holistic, collaborative, and reflexive ways of knowing.
- Ethnography has its origins in the fields of social/cultural anthropology in a colonial era. Multiple layers of critical conversation over time – along with the uptake of ethnography across a vast range of disciplines and settings – have led to transformations, innovations, and a valuable tradition of reflexivity.
- Ethnography has practical pros and cons, including the amount of time it takes, being able to be done by an individual researcher, not needing specialised equipment, and requiring adaptability and commitment.
- Today, ethnographic research is being used for a variety of purposes, with ongoing value in equipping researchers to ask human questions about human problems, challenge the taken-for-granted, and create ways to learn with (rather than just about) people.

RECOMMENDED FOR FURTHER READING

Coffey, A. (2018). *Doing ethnography*. London: SAGE Publications. Available from: https://doi.org/10.4135/9781526441874.

Madden, R. (2017). *Being ethnographic: a guide to the theory and practice of ethnography*. 2nd ed. London: SAGE Publications.

O'Reilly, K. (2012). *Ethnographic methods*. 2nd ed. Florence: Taylor & Francis Group.

Wolcott, H.F. (2008). *Ethnography: a way of seeing*. 2nd ed. California: AltaMira Press.

REFERENCES

Forsey, M.G. (2010). Ethnography as participant listening. *Ethnography*, 11(4), pp. 558–572.

Haraway, D.J. and Goodeve, T. (2018). *Modest_Witness@Second_Millennium. FemaleMan_Meets_OncoMouse: feminism and technoscience*. Routledge.

Miller, D. (2017). Anthropology is the discipline but the goal is ethnography. *HAU: Journal of Ethnographic Theory*, 7(1), pp. 27–31. Available from: https://doi.org/10.14318/hau7.1.006.

Pink, S. and Morgan, J. (2013). Short-term ethnography: intense routes to knowing. *Symbolic Interaction*, 36(3), pp. 351–361. Available from: https://doi.org/10.1002/symb.66.

Robben, A.C.G.M. and Sluka, J.A. (2012). *Ethnographic fieldwork: an anthropological reader*. 2nd ed. Chichester: Wiley-Blackwell (Blackwell anthologies in social and cultural anthropology, 9).

Shah, A. (2017). Ethnography? Participant observation, a potentially revolutionary praxis. *HAU: Journal of Ethnographic Theory*, 7(1), pp. 45–59. Available from: https://doi.org/10.14318/hau7.1.008.

Van Maanen, J. (2006). Ethnography then and now. *Qualitative Research in Organizations and Management*, 1(1), pp. 13–21. Available from: https://doi.org/10.1108/17465640610666615.

Wolcott, H.F. (2005). *The art of fieldwork*. Rowman Altamira.

Wolcott, H.F. (2008). *Ethnography: a way of seeing*. California: AltaMira Press. Available from: http://ebookcentral.proquest.com/lib/otago/detail.action?docID=1343764 [accessed 28 September 2022].

2

DESIGNING THE RESEARCH, DEFINING THE FIELD

Ethnography is a research methodology that has distinguished itself through the practice of **fieldwork**. Thus it is not surprising that 'the **field**' has been an important and influential concept. But ethnographers increasingly deal with a variety of different *types* of fields or with nontraditional approaches to fieldwork. By discussing and deconstructing the idea of the field, this chapter is both an introduction to ethnography in general – its approaches, core characteristics, and boundaries – and a practical guide for thinking through some of the first decisions you will have to make in designing an ethnographic research project. It does so via mapping out some of the ways the focus of ethnographic research has changed over time, charting shifts from colonial models of travelling to gather data in a remote 'village' to ethnography conducted in urban settings, 'at home', or using a multisited approach. It addresses different sorts of field, including organisations and institutions, and virtual sites. The chapter then shifts into practical questions about what type of engagement is required with a field, including how long to spend there, how to refine a topic or question to drive the research, and who will count as a 'participant'. It concludes with some important notes on safety and risk, along with a summary of recent conversations emphasising more flexible and inclusive approaches to fieldwork that provide another route to deconstructing traditional assumptions about the field while retaining a commitment to a specifically ethnographic approach.

DOI: 10.4324/9781003404880-2

WHERE (AND WHAT) IS A FIELD?

Ethnographers value the type of nuanced and **situated** social knowledge that is produced by 'being there'. But where is 'there'? In its most simple sense, the field is where you do your research. It is the real-world social setting in which you observe, immerse, and/or participate and where you can be co-present with your **participants** (Beaulieu 2010), in order to learn about their lived experiences and try to gain an insider perspective. However, though the term 'field' may conjure up the idea of a physical place, an ethnographic field isn't primarily defined in spatial or geographic terms. Instead, it is more typically about a *social* space: more a 'who' than a 'where'. With this in mind, an ethnographer might, for example, identify their field as a particular group or community, or might focus on an institution, organisation, or professional setting, or on a particular movement or subculture.

The next thing to know about the field is that it doesn't exist. Or at least, not as a natural, pre-existing object. Rather, it is socially constructed by the researcher as part of the process of identifying, defining, and delineating a focus for their study and through the activities they undertake and the ways they choose to represent it. Of course, a field may, at times, be something that can be recognised by others too – for example, a sports club, a rural town, or religious group all might exist on paper as well as existing in other people's minds. Even so, most social fields are fluid and porous, involving complex relationships between people, places, and social institutions, and it is typically the ethnographer who draws somewhat arbitrary boundaries around it by deciding in their own mind what (and who) they will include or exclude for the purposes of their project. As such, defining a field is less about recognising an actual boundary and more about focusing your own attention by putting practical and realistic limits around the scope of your inquiry, ideally with an openness to adapt later if you need to.

Importantly, for the people who are part of it, the field is a site of life as well as a site of research. As such, it remains a space of ethical, relational, and practical negotiations for the researcher, considering not only the conceptual parameters of your project but also your responsibilities to the other people involved (see also Chapter 3).

STORY BOX: The home is where it happens

"The home is a context where some of the most important things that happen are lived". For her ethnographic study of home laundry practices, Sarah Pink worked with twenty UK families. She started with in-depth interviews of each person. Then, she and her research team visited people's homes for a day or for a few hours over several days to be with them as they did laundry and other related tasks. She also asked people to invite her into these everyday spaces using film, getting participants to record themselves giving her video tours of their homes. Using the home as a site of ethnographic analysis allowed her to consider the practical activities that happen there and the way these are tuned to forms of sensory knowledge. What she learned from studying the "lived and continuously changing environment" of the home had wider implications for policy-making about health, energy use, and thus sustainability, housing design, urban planning, and other practical and applied topics *(Pink et al. 2015)*.

The terms '**field**' and '**fieldsite**' are also often used in slippery or interchangeable ways. While the 'field' is about the broader social space you are interested in, the 'fieldsite' is more likely to refer to the particular location/s in which you spend time, interact with people, talk, or listen. But while some fields will be strongly tied to a physical place, others won't at all, and will instead move locations or span multiple locations or be linked mostly to abstract or virtual spaces (such as an online platform, or a set of archives (see Chapter 5)). The way of recognising or approaching this has changed throughout the history of ethnographic research.

A BRIEF HISTORY OF THE FIELD

The language of the 'field' and 'fieldwork' recalls earlier ages of ethnographic research and, in particular, eras of colonial voyage and exploration (Massey 2003). Today, 'fieldwork' simply refers to original **empirical** work done in a **naturalistic** setting. This can take a huge variety of forms. Nonetheless, looking at changes to the idea of the ethnographic field over time provides its own window into how ethnographers aim to see the world.

Figure 2.1 Students shadowing a Forestry Commission woodland officer on their visit to a farm as part of a user research project in the UK.

Source: Photo credit: Dan Barwick and Becky Miller.

FROM THE VILLAGE TO THE CITY

Towards the end of the 1800s, amidst an era of colonial encounter, European-based scholars showed increasing interest in the studying the diversity of human social and cultural forms. Their methods for doing so had previously mostly involved 'armchair' anthropology, in which they examined secondhand accounts (by explorers, traders, missionaries, and folklorists) in order to posit theories about the social world. Some social scientists had also begun to join in European scientific voyages in order to conduct short 'surveys' with local populations (Robben and Sluka 2012). The introduction of long-term, immersive **fieldwork** in the early 1900s was quite a radical departure from either of these approaches. As Chapter 1 explained, and Chapter 4 will pick up, it facilitated the new method of **participant-observation,** and an approach much more focused on immersion and collaboration with participants. The two people

considered responsible for pioneering this approach were Franz
Boas and Bronislaw Malinowski.

Franz Boas was German born but trained and worked in the
USA, where he is seen as one of the founding fathers of cultural
anthropology. His research with First Nations people in North
America, beginning in the late 1800s, is seen as some of the earliest applications of ethnographic fieldwork. His work, especially at
the beginning, was based on shorter trips of perhaps a few weeks
or few months at a time, mostly because of limitations in funding.
Over the course of his career, Boas visited forty different sites
along the northwest coast of America, emphasising the need to
document the details as well as the contexts of the cultural practices he observed. He collaborated extensively with native informants to achieve this, including George Hunt, a Tlingit man who
worked with Boas as a guide, interpreter, and assistant for over
forty years and himself provided rich ethnographic notes and cultural collections on the language and rituals of the Kwakwa-
ka'wakw (Robben and Sluka 2012).

Starting slightly later, in the early 1900s, but more active in writing directly about his new ethnographic methods, was Bronislaw
Malinowski, a Polish-British anthropologist. Malinowski travelled
to the Trobriand Islands in 1914 to conduct research. His fieldwork
there was based on three visits spanning six years in total. His suggestion that ethnographers invest time periods as significant as these
in order to really immerse themselves in their field was quite radical. Malinowski's writing about his own such processes in Melanesia – interspersed with descriptions of the actual people and practices
he was documenting there – became some of the most famous ethnographic work ever published. It also started some of the stereo-
typical tropes of ethnographic fieldwork. In one of the opening
passages of his seminal book, *Argonauts of the Western Pacific*, he
describes the following arrival:

> Imagine yourself suddenly set down surround by all your gear,
> alone on a tropical beach close to a native village, while the
> launch or dingy which has brought you sails away out of site
> [. . .] Imagine further that you are a beginner, without previous
> experience, with nothing to guide you and no one to help you.
>
> (Malinowski 1922, p. 4)

Figure 2.2 Ramiro and Oscar, who are immigrants from Ecuador, playing music in Jean-Talon station in Montreal, Canada, photographed by ethnographer Nicholas Wees as part of his study of urban buskers, June 2016.

Source: Photo credit: Nick Wees (2016).

Accounts like this helped to entrench a romantic, heroic idea of the ethnographer as a person who takes arduous voyages far from home to live in a remote place. It also set up an expectation of the focus of ethnographic fieldwork being a small-scale, pre-industrial, and non-Western community, meaning ideas of 'the field' quickly became entangled with colonial imaginaries of 'the village'. However, the generation of students trained by Boas and Malinowski – and through whom ethnography became professionalised – were both witnesses to and leaders of some significant changes. Most notable was the adoption of ethnographic techniques to study larger-scale urbanised, industrial societies.

The scholars involved in establishing urban ethnography were based at the University of Chicago in the 1920s and 1930s, forming part of a movement that became known as the Chicago School. This included many sociologists and anthropologists who were interested in studying how urban societies are organised and how different places, institutions, or social forces associated with the city might be analysed in relation to issues like poverty or crime. While they drew on a mixture of quantitative and qualitative methods, they had a crucial role in promoting the long-term and immersive and fieldwork-based methods of ethnography (and, in particular, **participant-observation**) as a way to study everyday life in the city.

While ethnography has sometimes been described as the study of **culture,** ethnographers do not have to take this as a primary or sole focus. Indeed, developments in urban ethnography showed that people's identities, practices, norms, and social relationships are shaped not only by their cultural background but by association with other types of organisations and institutions as well. Some of these are default associations and some associations of choice. Some are fleeting and some are based on long-term commitments. An interest in all of these led to establishment of organisational ethnography and institutional ethnography, where the field is usually delineated around kind of legal-bureaucratic social unit such as a school, hospital, business, sports club, community organisation, and so on. The ability of ethnography to generate insights on workplace cultures and behaviours, managerial relations, emotional labour, health and safety, and so on made it of increasing interest to many professional groups for applied purposes too.

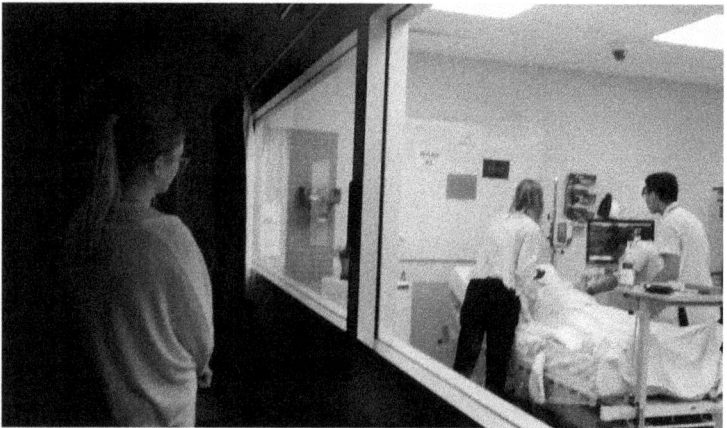

Figure 2.3 Medical students in a simulated clinical scenario undertaken as part of their training, being observed through one-way glass by other medical students (while the ethnographer observes both). Auckland, New Zealand, 2018.

Source: Image credit: Tanisha Jowsey.

STORY BOX: Moving with the media

For several years, Sahana Udupa spent time wherever journalists were working, both on and off the clock. As a media anthropologist conducting an ethnographic study of news cultures in the aspiring 'global' city of Bangalore, India, she carried out her observations in the newsrooms of several major newspapers, sitting through meetings, conducting interviews, interacting, and chatting with journalists on the newsroom floor. She examined the news texts they produced. She also examined online spaces and texts, such as the informal blogs in which journalists exchanged gossip and expressed discontent with their managers and organisations, typically under pseudonyms. Udupa had many conversations with journalists outside of their official locations of work, spending time at professional associations, private homes, and so on. In sum, she describes having "moved with the news narratives around the city, following the people who made, remade and enacted them in a changing urban milieu" (Udupa 2015).

FROM FAR AWAY TO AT HOME

While the earliest ethnographies portrayed 'the field' as geographically and culturally separate from the ethnographer's own social worlds, the Chicago School disrupted the image of the ethnographer as someone who 'goes somewhere' by ushering in the possibility that ethnographers could conduct research in places in which they were already embedded. This also disrupted the idea of the ethnographer as someone who starts as an outsider. Valuable insights began to come from insider ethnographers analysing their own familiar settings or communities. This was supported by turns challenging broader assumptions the role of **objectivity** vs **subjectivity** in ethnographic research (see Chapter 4). What it meant to do ethnography 'at home' had to be critically deconstructed too, however. Home can be a complex and taken-for-granted category, easily applied in a way that masks differences between people or groups in a given setting (Madden 2010, p. 10). *Are you at home anywhere in your town or city? Or only within specific groups or communities there? Is home defined by how you feel or by how other people feel about you?* With this in mind, ethnographers had to reflect on not only what/who constituted their field but also on their **positionality** *in* that field, recognising the partiality of categories like 'insider' or 'outsider' (see Chapter 3).

The work of the Chicago School tended to focus on groups that were, in some ways, outsiders in their own society, including ethnic minorities (e.g. migrant communities), places or people who were marginalised or stigmatised in some way (e.g. strip joints, hobos, professional thieves), closed institutions or geographic conclaves (e.g. retirement homes, Jewish ghettos), or specific subcultures (e.g. taxi-dance halls, boy gangs) (Robben and Sluka 2012). This focus had a valuable role in highlighting the experiences and agencies of groups who might not traditionally have a voice among academic institutions or policy-makers. In many ways this has continued in contemporary ethnographic practice. At the same time, within a few decades, other scholars were challenging the assumptions and the power inequalities involved in this and suggesting ethnographers should study majority cultures and powerful people as well (see also Chapter 3).

FROM BOUNDED SITES TO THE MULTISITED AND VIRTUAL

Ethnographic research began at a time when 'culture' was often conceived as a static, bounded thing attached to a single group in a single place. This shaped ideas about the ethnographic field. Some ethnographers were strongly motivated by a view of non-Western cultures and traditions as fragile or vulnerable to being lost once in contact with outside forces such as modernisation or globalisation. Later, the term 'salvage ethnography' was used to critique the type of work that focused on documenting non-Western cultures before they 'disappeared'. But while there has indeed been a documented loss of languages and destruction of both tangible and intangible cultural heritage through these historical forces, it also isn't quite so simple as that (Eriksen 2015). Eventually, ethnographers had to grapple with the idea of the global in a more nuanced way, recognising that many individuals and communities had relational networks, identities, and practices that extended transnationally, and through which they intentionally pursued their own goals and agendas. Amidst lives shaped by migration and movement, global economic systems, and the circulation of mass media and popular culture (Abu-Lughod 2000, p. 263), change does not always or only represent loss, but leads to variety of different, new, hybridised, or 'glocal' cultural forms. This contributed to new understandings of cultures as porous, fluid, and adaptive. All of this helped, over time, with a deconstruction of the idea of the bounded field, and with recognising the field itself as a "colonized concept" (Qamar 2020). But it also made the task of ethnography more complicated.

American anthropologist George Marcus responded through his writing on the "multi-sited research imaginary", suggesting a new way to think about the ethnographic field. He proposed **multisited** ethnography as a type of ethnography which "move[s] out from the single sites and local situations of conventional ethnographic research designs to examine the circulation of cultural meanings, objects and identities in diffuse timespace" (Marcus 1995, p. 96). Today, many ethnographers design their projects in a way that recognises and attends to more than one fieldsite for these reasons. Importantly, multisited ethnographies are not just about getting *more* data. Nor are they necessarily about comparing different sites (though they can do this usefully sometimes too). Instead, they respond to a desire to analyse social phenomena as they move, adapt, and connect people in multiple ways and across multiple scales.

STORY BOX: How to sing about Babylon in Japan

Marvin Sterling is a Black American anthropologist whose work has explored contemporary Japanese engagement with Jamaican culture. He did this over two years of fieldwork, between 1998 and 2000, during which he focused largely on the Tokyo metropolitan area but also travelled around the country "wherever I could find people, places, and events related to dancehall, roots reggae, and Rastafari" (Sterling 2010). Several trips to Jamaica provided additional context for his work on "global imagination of Blackness". This helped him study the "unmooring, transnational" circulation of pop-cultural artifacts and practices by looking at how Caribbean pop-cultural forms were re-localised to create new, different, but still meaningful experiences for his Japanese participants: for example, considering how they might connect with Bob Dylan's songs about 'Babylon', given this is a very specific political metaphor for Western colonisation, and Japan has a totally different relationship to colonisation and the West.

Digital and internet technologies have been part of reconfiguring the enactment of social life at these global scales. They have become entangled with people's ways of being in the world and being with each other, providing virtual 'spaces' or 'places' where people interact, educate, discuss, entertain, debate, care, and more. In other words, they have become fields where social life is lived out. As such, digital platforms have also become valid sites for ethnographic research. The practice of ethnography in these sorts of fields has been variously referred to as digital ethnography, cyberethnography, virtual ethnography, ethnography online, or netnography (Miller 2018). Identifying an online space as a fieldsite can be challenging, since they tend to be spaces "with fluid boundaries and fragmented information" (Wang and Liu 2021, p. 979). *How does the ethnographer know what (or who) is 'in' or 'out' of the digital field when they are by their very nature networked?* In a way, however, this is not so different to other fieldsites, and digital infrastructures provide their own clues as to how people are connected to each other and/or to particular groups, pages, or communities.

STORY BOX: Arriving on the shores of the virtual

Tom Boellstorff, a US-based anthropologist, had previously travelled to fieldsites in Indonesia to research sexuality using ethnographic methods. When he started a project on the virtual platform called *Second Life* in June 2004, he stepped into an entirely different sort of field. What followed was two and a half years of participant-observation and interviewing, but through his virtual presence as an avatar, in the virtual home and office he built and named *'Ethnographia'* (Boellstorff 2008, p. 4). This work was one of the earliest examples of applying ethnographic techniques to an entirely virtual fieldsite (7), and Boellstorff was adamant that the "challenges and joys" of his experiences as a researcher in *Second Life* closely resembled his more traditional fieldwork in Indonesia (70). To highlight this, he opens his book with a comparison between his 'initiation' into *Second Life* and Bronislaw Malinowski's experience of arriving at his Melanesian fieldsite. Continuing to playfully reference earlier ethnographic tropes, he titled his book *Coming of Age in Second Life* (2008) to allude to Margaret Mead's famous work on *Coming of Age in Samoa* (1928).

Figure 2.4 Tom Boellstorff's avatar 'flying' across the landscape of his *Second Life* fieldsite.

Source: Image credit: Tom Boellstorff.

Boellstorff has advocated that participating in online communities in the same way other users do is a valid form of fieldwork – a way to study these social worlds "in their own terms" while still applying an **ethnographic lens** in order to identify the infrastructures, relationships, and norms of that virtual world (2008, p. 4). However, no one lives totally online, and not only do social fields often span both offline and online social worlds, but ways of engaging in the digital remain embedded in material technologies and embodied practices. For this reason, Daniel Miller has argued that continuing to focus on the broader context of people's lives remains useful to understanding digital practices in the manner of 'holistic ethnography' (Miller 2018, p. 1).

WHAT DO YOU WANT TO KNOW?

All researchers have to narrow down, at some point, what they are planning to research. For ethnographers, who are interested in situated knowledge, this usually also involves identifying a field. In earlier eras, there was an expectation that ethnographers were gathering data in order to document this field in its entirety. This expectation often shaped the choice of fieldwork locations, with a bias towards studying communities with "a population no larger than what a lone ethnographer might come to know in the course of a year or so" (Wolcott 2008, p. 23). Today, the idea of documenting a whole culture is almost laughable and not something we typically aspire to. Instead, ethnographers are encouraged to have a particular topic, problem, or question they are interested in; in other words, some particular *aspect* of the field that they would like to find out about. This might be a question or problem of significance to the people in the field, or it might be based on a more theoretical question that emerges from scholarly dialogue. Often, it is a mixture of both.

> **STORY BOX: Questioning adolescence**
>
> American cultural anthropologist Margaret Mead can be credited as the first major contributor to 'problem-based ethnography'. Margaret Mead was trained at the Chicago School under Franz Boas and Ruth Benedict. Setting out for her first fieldwork expedition in Samoa in 1925, Mead was very interested in studying adolescence. In particular, she was driven by the question of whether the teenage behavioural patterns she observed in American society, and which were assumed by many to be natural or biologically driven, were actually universal. She sought to gather empirical data through observing adolescent behaviour in a very different cultural setting in order to compare it with her own more familiar settings. She chose the Pacific nation of Samoa, and from her observations and conversations there over about nine months of fieldwork, she concluded that a lot of ideas and norms around adolescence were, in fact, socially constructed and culturally specific.

ETHNOGRAPHIC RESEARCH QUESTIONS

Research questions can be "orienting" (Wolcott 2008, p. 74). Even if you have identified a field and a general topic, a research question can help explain what it is you want to know *about* that topic. The types of questions that are useful for ethnographers are sometimes quite different to those that might be used in a more **positivist** type of study. Scientific studies often have a hypothesis – a specific idea that they want to test in order to prove or disprove it. This is described as **deductive** research. Ethnography more typically employs an **inductive** approach. That means that although we might start with a question, we aren't going in with a specific pre-formed hypothesis to test. Instead, our questions identify an area of exploration with the expectation that we will develop more specific theories based on the research process itself.

As a **qualitative** form of inquiry, ethnographic research questions work best when they are open-ended, inviting a descriptive, nuanced account rather than a 'yes/no' answer. Like any research question, you should think carefully about its scope and focus, assessing what you can actually answer within the practical parameters of your project and bearing in mind factors like time, money, the specific methods of data collection you are using, and your access to (or within) the fieldsite. Ethnographic research questions should also be transparent about the situated nature of the knowledge they are seeking. I personally prefer ethnographic research questions that state the field within the question itself to make clear that the study is setting out to make a contribution to some broader conversation by gathering data related to *this* specific group in *this* specific place and time.

WHO SETS THE COURSE?

A researcher's choice of fieldsite, topic, or question doesn't come out of nowhere. These decisions can be shaped by a variety of factors ranging from the personal and interpersonal, to the institutional and practical, to wider political or ideological tides.

Our choice of topic, focus, and field may relate closely to the people we are, our background and life experiences, and our values, as well as our social relationships. Ethnography requires (and is enriched by) careful reflexivity around this. This is not in order to mask, hide, or reduce our personal presence or moral commitments but rather to factor them in actively and transparently. *Which fields is the researcher already connected to? How have they come to take on the role of researcher? Why here and why now?* In ethnographic research, it is appropriate and expected that the participants will also shape the direction of the research in a variety of ways (see

Chapter 3). This is particularly sought and prioritised in ethnographic research with a collaborative or activist framework. Simultaneously, however, institutional structures and professional relationships outside of the field can shape the direction of our research. This can be built into systems such as academic supervision or project management, that shape our decisions directly. But our interests and approaches may also be shaped by those we engage with indirectly, including scholars whose work we are recommended or taught, or read in early stages of our own planning. As such, it is worth continuing to cultivate a contextual and critical awareness of research institutions themselves. *What are the politics or trends shaping what work gets published, taught, promoted, cited, commissioned, or funded? What can my research contribute to existing conversations, or what can it offer that is new or different?*

Almost all research is dependent on some or another institutional body to approve it or fund it or both. Often, some of the first writing an ethnographer does about their project, is to write research proposals or funding applications. The trends of funding for particular topics or particular fields can shift over time or may be skewed because of the geographic location of the institutions that are training and funding ethnographers. Funding committees are interested in projects that seems realistic and achievable, so those that seem likely to face challenges around time, distance, or access might be less likely to get funding. Ethics approval applications are another type of writing undertaken early in the process, and these, too, can shape how the research is designed. Researchers may shy away from even pitching a project in which the field is mainly vulnerable people (e.g. children, incarcerated people, those with mental illnesses or disabilities) or the topic is particularly sensitive or controversial lest this slow the approval process, especially if they have a strict timeline to meet. But this can be a real loss, as ethnographers do, in fact, have tools that allow them to contribute thoughtfully and sensitively in these sorts of areas (see Chapter 3).

ADAPTING AS YOU GO

According to Paul Atkinson, "Even less than other forms of social research, the course of ethnographic work cannot be predetermined, all problems anticipated, and ready-made strategies made available for dealing with them" (2007, p. 20). He hastens to add that doesn't mean there is no need to prepare or that we should take a haphazard approach, especially given our ethical commitments to the real people and communities that comprise our fields. But there *is* a need for continuously rethinking one's focus and approach. Once you start your project, some pathways may unexpectedly open, and others may unexpectedly close. When this happens, you will ideally have enough freedom from your institutions, funders, and supervisors to adjust and refocus. Good fieldwork is, in fact, *supposed* to change the ethnographer's questions, goals, or methods in a process Anna Tsing calls "ethnographic learning" (2011). The encouraging thing is that the ethnographic process will never leave you with *nothing* to study. It may just mean that sometimes, you aren't studying exactly what you thought you would.

HOW LONG WILL YOU SPEND IN THE FIELD?

Traditionally, ethnographic research has been characterised not only by fieldwork but by *long-term* fieldwork. The purpose of this was to facilitate the researcher's immersion in the field and access to insider perspectives (see Chapter 4). This goal of spending months or years in a particular fieldsite was often seen as the mark of a successful and thorough ethnographic study. But fieldwork visits of this long are not always practical for contemporary researchers, nor may a longer-term approach suit contexts in which insights are needed urgently. For all these reasons, recent decades have seen innovations around applying ethnographic principles in projects that have shorter time frames. In public health research, for example, 'rapid assessment procedures' and 'rapid ethnographic assessment' were pioneered the 1980s based on a need to contribute to programmes of disease management quickly. Other approaches such as 'rapid ethnography' or 'focused ethnography' emerged in the early 2000s with similar aims (Pink and Morgan 2013). Short-term approaches have been criticised as risking being superficial in their analysis, but, as Pink and Morgan suggest, they may be able to compensate for the short period with data-intensive approaches that can be analysed in detail later (2013). While longer projects typically have relied on time for building rapport, hanging out, and 'waiting for things to happen', rapid projects rely on getting to the thick of it immediately. Shorter-term projects are also less likely to focus on **theory**. At the same time, they may require even more focused research questions to shape an efficient approach.

ACCESSING THE FIELD

Since a fieldsite is a social space, not just a geographic place, access is about more than turning up. Rather, it is a nuanced issue and one that ethnographers will have to think carefully about. Cunliffe and Alcadipani suggest that access has two levels (2016, p. 537). On one level – that of 'primary access' – it can be about obtaining formal permission to enter, visit, or become part of a location, organisation, institution, or community in order to undertake research there. This can be a formal process, involving everything from visas to research permissions from organisations to agreements with elders. Sometimes, it isn't clear at the start what these processes should even be, or who to approach first, and so it really can take time to sort out.

STORY BOX: The ethnographer, in the lab, with a gun

The work undertaken by forensic scientists is typically "buried in windowless basements of municipal buildings and in far-flung warehouses", but it is essential to the pursuit of criminal justice in the USA. Beth Bechky is an organisational ethnographer. After a fortuitous opportunity to tour a crime lab, she became curious about the mixture of "mundane and demanding" daily work the scientists there were undertaking and the scrutiny they often encountered in the court system. When she expressed interest in starting an ethnographic study, the directors of the lab asked her to write a proposal and then left her "anxiously awaiting their permission" for over a year. Eventually, every lab director she had approached declined. However, four years later, an opportunity unexpectedly opened up, and Bechky was able to spend eighteen months moving between four applied science units inside the Metropolitan County Crime Lab, doing participant-observation three days a week for between three and six months in each. She couldn't participate in the work itself (so as not to pollute chain of evidence), and there was some emphatic back and forth over whether she was allowed in the firearms lab or not. Despite this, she was able to observe, attend meetings and training sessions, and conduct interviews, leading to a successful book about the pressures present in this professional world (Bechky 2020).

On another level – that of 'secondary access' – access is about building relationships in order to gain access to people, spaces, experiences, and knowledge *within* the fieldsite (see also Chapter 3). This is mostly about informal and social access, but can be just as challenging and time-consuming to negotiate, since it relates to the internal politics of that social world as well as the researcher's own identity and positionality within it.

WHO ARE YOUR RESEARCH PARTICIPANTS?

Earlier in this chapter, I suggested the ethnographic field is more of a 'who' than a 'where'. Despite this, defining your field is not the same as defining who your participants are. Institutional processes for granting ethics approval – sometimes described as 'procedural ethics' – typically focus on formalising the designation of participant by requiring researchers to get anyone they are collecting data from or about to sign a consent form. But these procedures were often designed with a different sort of research in mind and may sometimes feel like a poor fit for ethnographic methods. Since ethnographers will often interact with many people in their field over time, but in a variety of different ways, they will need to consider who, for formal purposes, will count as a participant. Think of an ethnographer who is studying a rural farmers market, an online investors platform, the doctors at an urban hospital, or a troupe of street performers. *When you are conducting participant-observation, is everyone in the vicinity considered a participant? If so, how can you ensure they are all informed about your presence and purpose there, and how do you get consent from them all? Or is it only the people you conduct recorded interviews with or gather data directly from?* It is important to feel settled in how you will approach this in a commonsense but ethically appropriate manner.

The same applications often ask researchers to state their recruitment method, and '**sampling** criteria' – that is, the parameters around which they will seek, advertise to, or invite participants, including any specific inclusion or exclusion criteria. There are different approaches to sampling that suit different types of research. Ethnographers often take a 'purposive' approach by approaching those they think will offer the most insight on their topic. Alternatively, they may rely on word of mouth, or snowball sampling, in which one participant might recommend or introduce another or in which the connections are formed over time. Overall, while ethnographers don't tend to emphasise sampling in the same way as some other fields, many of the underlying questions remain useful: *Who will you speak to or spend time with? What knowledge claims can you make based on insights from these people (and not others)? Who might be present in your field but typically not acknowledged? What diverse or conflicting perspectives or experiences may different people in the same fieldsite have?* Remaining flexible to engage with people you hadn't planned to originally or who weren't originally counted as part of your field at all can be extremely valuable.

NONHUMAN SOCIAL ACTORS

For some ethnographic projects, it is also important to not be limited by assumptions about who the social actors in a field might be. Several important school of thought in social research, philosophy, and theory – including within science and technology studies (STS)

Figure 2.5 Sophie Chao participating in cooking sago in a palm grove, with Marind community members, in Merauke, West Papua.

Source: Photo credit: Selly (2016).

and feminist new materialism, for example – recognise objects, technologies, and places, as having their own social agencies. This idea is not new at all, from the perspective of many Indigenous **worldviews**, where "all things are living", and the knowledge, memory, stories, wisdom of people groups are carried by the land itself and by different material substances and forms (Smith 2020).

STORY BOX: Acknowledging a murderous plant

Sophie Chao worked for an international Indigenous rights organisation for many years. But she felt drawn to go deeper in studying people's experiences of the "boundless devastation and disciplined monotony of industrial monocrops" in rural West Papua. Over three years, she spent time in three villages belonging to the Marind people on the upper reaches of the Bian river. Here, a $5 billion agribusiness had been launched, and Chao wanted to learn how the Marind engaged with their changing landscapes. She learned that they didn't see the oil palm plant as simply an object of human exploitation but rather as a "willful entity" that was transforming the ecosystem and their lives (5); viewing it as a "rapacious and foreign plant being'" (4) that was killing their beloved sago plants, choking their rivers, and bleeding their land (Chao 2022).

Recognising nonhuman social actors can be particularly important for some fieldsites and/or for certain types of ethnographic research, such as multispecies ethnography. This type of ethnography is interested in both human and nonhuman species (sometimes also called 'other-than-human' or 'more-than-human' social actors), often with a focus on their ways of interacting in shared spaces. To explore this, ethnographers may seek innovative or experimental ways to approach animals, plants, fungi, and so on, as 'participants' in their own right – though the question of how to apply ethical principles to these interactions is not straightforward. At the very least, this ensures their presence and agency can be acknowledged.

Equally, in some settings, ghosts, deities, or ancestors may have names, identities, and myriad agencies through which they shape human lives. Can they also be participants? Taking this idea seriously might, in certain contexts, be a way of challenging or decentering some of the assumptions in the Western colonial paradigm of research (Lowe et al. 2020), a pathway to working more deeply with or from Indigenous **epistemologies** as a useful part of a **decolonising** response.

THE SELF AS PARTICIPANT: ABOUT AUTOETHNOGRAPHY

Another form of ethnography that complicates the definition of the 'participant' is **autoethnography**. Autoethnography means that the researcher is using themselves – their own experiences or practices – as their focus, their field, and their main source of data. It first appeared in discussions of ethnography in the mid-1900s amidst conversations in feminist ethnography that emphasised personal experiences as a valid window into social analysis. It is important that auto*ethnographic* writing goes beyond the auto*biographical* by analysing, deconstructing, and theorising as well as describing (Delamont 2009). Dutta argues that autoethnography in particular

Figure 2.6 Susan Wardell (the author) and her young son examining a beehive in their suburban backyard as part of an autoethnographic project on hobbyist beekeeping and moral life in the Anthropocene.

Source: Ōtepoti Dunedin, New Zealand, 2022.

can work as a decolonising tool by "radicalizing knowledge production" and unsettling the privilege of the scholar as a figure traditionally embedded in colonial, Western-based institutions and methods of knowledge production (Dutta 2018).

There is no single definition of or approach to autoethnography. While overarchingly, it is "an attempt to systematically and self-consciously write oneself into cultural analyses, situating one's perspectives and experiences within larger cultural and social frameworks" (Ellis et al. 2011, p. 1), it is not always clear what might differentiate this from the wider paradigms of ethnography, since all ethnographers are expected to provide a degree of reflexivity. Some people have defined autoethnography as a form of research in which the author's experiences are the *sole* focus of analysis. Yet other projects labelled as autoethnographic may be more about analysing a social field the researcher is part of in a way that still involves other participants, often then overlapping with the approach of '**insider ethnography**'. Still other people have *resisted* having their work labelled as autoethnographic just because they share some personal features or experiences with their participants.

HEALTH AND SAFETY IN THE FIELD

Fieldwork often creates both memorable and challenging experiences for the ethnographer. It may also generate a variety of risks to health, safety, and well-being. Older attitudes to fieldwork sometimes emphasised suffering or struggle in the field as a sort of 'rite of passage' for an ethnographer, but this is an outdated and dangerous approach. Contemporary practice acknowledges that ethnographers in every type of setting require and deserve good support structures in order to proceed with fieldwork in a safe and successful way.

WELLBEING, CULTURE SHOCK, AND OVERWHELM

For some ethnographers, fieldwork means travelling to a new place or engaging in a new culture. For others, it may mean entering a new community, engaging with a new organisation, joining a new digital space, or partaking in new activities. The idea of 'culture shock' is sometimes used to describe how an unfamiliar social or cultural setting can be experienced in a distressing way, often at a bodily level. This can be about subtle and sometimes quite mundane things involved with the embodied rhythms of everyday life or adjustments to different social norms or expectations. The researcher is usually undertaking fieldwork based on a good deal of planning, preparation, and work – committed to spending a fair bit of time in this setting – and so these initially unpleasant feelings can be hard acknowledge. Missing the familiar is allowed. Being confused, frustrated, or tired is also allowed. There is no degree of professionalism that reduces these sorts of feelings. While it is important to keep an eye on basic thresholds of comfort and personal safety (as discussed next), it is also good to remember that at least some of these feelings may reduce over time. Building in time to settle in can help. Splitting up fieldwork periods to have a break in the middle can also be beneficial both personally and for **iterative** analytical processes (see Chapter 7).

Even if a field is familiar to us already, we take on both new perspectives and new responsibilities when we put on the ethnographer's 'hat'. Moving into a new role (i.e. as a researcher) in a space in which we are already intimately embedded can involve its own version of awkwardness, stress, or strain. It may shift existing social relationships or create competing commitments (see Chapter 3). The process of **defamiliarisng** (see Chapter 4) may generate new and sometimes discomforting forms of awareness of people or practices that are entangled with the ethnographer's personal life and identity. There also may be less opportunity to step away or take a break from research when research and life are overtly one. Ethnographers doing this type of work need support and self-compassion too.

DANGER AND RISK

Ethnographers may engage in all sorts of activities in their fieldsites and as part of participant-observation (see also Chapter 4). Ethics approval processes typically focus on checking whether the research plan is safe for participants, as is indeed crucial. But there is also a need to take seriously the safety of the ethnographer too.

Figure 2.7 A morning ride with a cowboy crew to bring cow–calf pairs back down to lower pastures for the winter, as part of Andrea Petitt's year-long 'horseback' ethnography in the Colorado Rocky Mountains, in 2019.

Source: Photo credit: Andrea Petitt

Dangers in the field can take many forms. In a small number of tragic cases, ethnographic research has resulted in the death of the fieldworker themselves or of someone accompanying them. Serious accidents and injuries are possibilities. Various forms of social, emotional, or psychological harm have to be considered. In some settings, there may be legal risks, such as being sued or being incarcerated, or financial risks, such as loss of property, theft, or scamming. There is also the risk of sexual harassment. As Markowitz and Ashkenzi point out, ethnographers have often joked about fending off unwanted romantic or sexual advances (1999), but this needs to be addressed seriously, as sexual assault can occur during fieldwork just as anywhere else. Risks are not evenly distributed between places or people, since, as Lee (1995) explains, these risks may be environmental, social, or biological hazards (related to the particular place or setting or to the nature of the topic being studied) but may also relate to the positionality (for example, age or gender) of the researcher.

Health and safety protocols for researchers vary widely in their scope and application but also tend to follow generic templates that are unlikely to capture all of the salient factors specific to your own setting. Ethnographers sometimes gain more detailed information about potential risks in the field through "corridor talk" (Lee 1995) – that is, the informal discussions people share behind the scenes. In some places, there may be a culture of flippancy, denial, or machoism around risks of fieldwork. But the myth of the lone fieldworker is just that . . . a myth. Supportive relationships both in and beyond the field are key, as is thoughtful planning and open communication. With this in mind, here are some suggestions about what to consider when selecting a field and planning for fieldwork:

- Research the fieldsite/s thoroughly ahead of time through both formal and informal channels.
- Engage with whatever health and safety processes and supports your institution has set up. Work out which people handle this, and be in contact with them about your specific project well in advance to discuss.
- Engage with any additional health and safety protocols that might exist in the fieldsite itself, for example, workplace/site training or risk assessment procedures.
- Check in with yourself regularly about how you are feeling. Take this seriously. Notice patterns or changes over time in your

physical, emotional, or mental state and/or the behaviours of those around you, and trust your instincts when something feels off.

- Keep open channels of communication with supervisors, mentors, or managers, as well as informal support networks of family and friends, including those who are outside of your fieldsite. Let them provide checks and balances through giving fresh perspectives on complex situations you have become immersed in.
- Identify supportive relationships and safe people *within* the field.
- Have a backup plan for worst-case scenarios, including a way to leave the field quickly if you need to.

Perhaps most importantly, know you are allowed to make decisions (to adapt the research plan, cease particular engagements, leave situations, or even end the project) for reasons of personal safety and well-being. Even if you are facing the possibility of offending people or losing research opportunities, it is okay to choose to protect your own health and well-being.

STORY BOX: Responding to rape jokes

Cari Tusing conducted ethnographic fieldwork in Paraguay. Her goal was to study how some of the new systems of collective land titles were affecting people in the Indigenous Guarana and campesino communities and also to include perspectives of local elites. However, her ability to move through this field and collect data safely was shaped by her race and gender. She explains that "My whiteness facilitated access to elite ranchers, while my gender exposed me to rape jokes and prying inquiries about my sexual availability".

Tusing developed her own strategies in response. Relationships of trust and dependency were key to this, including hosts who housed, cared for, and 'claimed' her. For example, when asked where she slept by some male ranchers – a question akin to a sexual advance – Tusing was able to explain that she 'slept' with her campesina mother. This 'mother', her host, was a woman her own age but who took on a kin role that was recognised by the community. This meant they could see Tusing as being part of and protected by that family unit. When not already accompanied by male members of her host family, Tusing also made the decision to hire a male field assistant to go with her when visiting and interviewing the ranchers (Tusing 2023).

As ethnographers, we may feel that there is a lot at stake around the success of our project. We may struggle to report when things go wrong. There can be internalised victim blaming, where we worry about being judged for not taking more measures to mitigate risk. This is especially so for stigmatised types of harm, such as sexual assault. But things can go wrong for anyone, for any number of reasons, even when the researcher does everything right. If something does go wrong, I encourage you to share quickly and fully with the people that can get you the support you need, usually those at your home institution. If you are not listened to immediately, keep trying, or try approaching different people, including going up the chain if needed. Once things are safely resolved, or when you feel able to do so, I also encourage you to share openly about the experience so that these conversations can be normalised.

BEYOND FIELDWORK AND THE FIELD

In reality, fieldwork is only one part of the research process. Talking with people in settings outside of the field, reading scholarly literature, engaging with popular texts, spending time with archives, and many other activities may also be rich and significant parts of ethnographic research. In fact, some ethnographic studies won't involve fieldwork at all or will involve ways of engaging with the people at the centre of our research that look dramatically different to 'traditional' ethnography.

Contemporary ethnographic work has been shaped by compounding disasters, pressures, and risks in the wider world. The COVID-19 pandemic made this particularly stark. At the same time, ethnographers' lives and practices continue to be shaped by overlapping personal and professional commitments. Given conditions that make traditional models of fieldwork challenging or impractical, calls for "flexible, care-driven research methods" have emerged strongly and in a manner that also involves deconstructing the 'normative' expectations of fieldwork and the field (Rivera-Gonzalez et al. 2022, p. 291). A recent and rich contribution has the writing on patchwork ethnography (Günel et al. 2020). This framework argues that limitations and constraints can open doorways to new possibilities for research. It recognises current shifts in practice – such as a move towards online methods, shorter-term field visits, and other innovations – and asks ethnographers to honestly and reflexively respond to the implications of this across four main areas, including:

- Reconceptualising relationship with home and field
- Accommodating news ways of 'being there'
- Collecting data in fragmented and patchworked ways
- Rethinking a linear temporalisation of the research process.

These authors ask how ethnographic principles might be applied with rigor and integrity within these adapted approaches, suggesting overarchingly that the core ethnographic commitment to long-term engagement and contextualised knowledge can still be taken as commitments for patchwork ethnography (Günel et al. 2020). This, at the same time, enables ethnographers to respond to older models of fieldwork that involve elitist and exclusionary assumptions about the positionality, mobility, or resources of the ethnographer by validating forms of ethnographic research that are more accessible, more realistic, and 'kinder' for a variety of researchers in a variety of different contexts, while still holding people accountable for **reflexivity** around how they are working to construct ethnographic knowledge.

CHAPTER SUMMARY: KEY POINTS

- Ethnographic research has traditionally been defined by fieldwork, with a focus on engaging with people in real-world settings in order to learn about their everyday lives.
- A 'field' is less about a geographic location and more about a social space the researcher will spend time immersed in, and trying to understand. A field also doesn't exist as a natural object but is constructed by the ethnographer as part of defining the focus and bounds of their research.
- Assumptions about the field, and the ethnographer's relationship to it, have been challenged over time, including work to unpack colonial stereotypes, and to respond to globalisation. Ethnographers increasingly work with many different types of field, including multisited projects, virtual sites, or 'patchwork' approaches.
- Along with defining a fieldsite, ethnographers need to identify a topic and/or research question. They need to think about who they

will try to connect with in the field (i.e. their participants), while remaining open to learn, adapt, and adjust as the project develops.

- Fieldwork can be a big and sometimes challenging experience no matter where it happens. It involves potential danger and risk. Ethnographers should take seriously their own safety and well-being in the field alongside their ethical commitments to participants.

RECOMMENDED FOR FURTHER READING

Beaulieu, A. (2010). From co-location to co-presence: shifts in the use of ethnography for the study of knowledge. *Social Studies of Science*, 40, pp. 453–470.

Günel, G., Varma, S. and Watanabe, C. (2020). *A manifesto for patchwork ethnography, society for cultural anthropology*. Available from: https://culanth.org/fieldsights/a-manifesto-for-patchwork-ethnography.

Massey, D. (2003). Imagining the field. In: Pryke, M., ed. *Using social theory: thinking through research*. London: Sage Publications.

Robben, A.C.G.M. and Sluka, J.A. (2012). *Ethnographic fieldwork: an anthropological reader*. 2nd ed. New York: John Wiley & Sons.

REFERENCES

Abu-Lughod, L. (2000). Locating ethnography. *Ethnography*, 1(2), pp. 261–267.

Atkinson, P. (2007). *Ethnography: principles in practice*. London, UK: Routledge. Available from: http://ebookcentral.proquest.com/lib/otago/detail.action?docID=308687.

Beaulieu, A. (2010). From co-location to co-presence: shifts in the use of ethnography for the study of knowledge. *Social Studies of Science*, 40(3), pp. 453–470.

Bechky, B.A. (2020). Appendix: case notes on an ethnography of a crime laboratory. In: *Blood, powder, and residue*. Princeton University Press (How crime labs translate evidence into proof), pp. 189–198. Available from: https://doi.org/10.2307/j.ctv13qfvpn.12.

Boellstorff, T. (2008). *Coming of age in second life: an anthropologist explores the virtually human*. revised ed. Princeton: Princeton University Press. Available from: https://doi.org/10.2307/j.ctvc77h1s.

Chao, S. (2022). *In the shadow of the palms: more-than-human becomings in West Papua*. Durham: Duke University Press. Available from: https://doi.org/10.2307/j.ctv2j86bm4.

Cunliffe, A.L. and Alcadipani, R. (2016). The politics of access in fieldwork: immersion, backstage dramas, and deception. *Organizational Research Methods*, 19(4), pp. 535–561. Available from: https://doi.org/10.1177/1094428116639134.

Delamont, S. (2009). The only honest thing: autoethnography, reflexivity and small crises in fieldwork. *Ethnography & Education*, 4(1), pp. 51–63. Available from: https://doi.org/10.1080/17457820802703507.

Dutta, M.J. (2018). Autoethnography as decolonization, decolonizing autoethnography: resisting to build our homes. *Cultural Studies ↔ Critical Methodologies*, 18(1), 94–96. https://doi.org/10.1177/1532708617735637.

Ellis, C., Adams, T.E. and Bochner, A.P. (2011). Autoethnography: an overview. *Forum Qualitative Sozialforschung/Forum: Qualitative Social Research*, 12(1). Available from: https://doi.org/10.17169/fqs-12.1.1589.

Eriksen, T.H. (2015). Anthropology and the paradoxes of globalisation. In: *Small places, large issues*. 4th ed. Pluto Press (An introduction to social and cultural anthropology (4th ed)), pp. 367–390. Available from: https://doi.org/10.2307/j.ctt183p184.23.

Günel, G., Varma, S. and Watanabe, C. (2020). *A manifesto for patchwork ethnography, society for cultural anthropology*. Available from: https://culanth.org/fieldsights/a-manifesto-for-patchwork-ethnography [accessed 9 June 2023].

Lee, R.M. (1995). *Dangerous fieldwork*. London: SAGE Publications. Available from: https://doi.org/10.4135/9781412983839.

Lowe, S.J., George, L. and Deger, J. (2020). A deeper deep listening: doing pre-ethics fieldwork in Aotearoa New Zealand. In: George, L., Tauri, J. and Te Ata o Tu MacDonald, L., eds. *Indigenous research ethics: claiming research sovereignty beyond deficit and the colonial legacy*. Bingley: Emerald Publishing Limited (Advances in Research Ethics and Integrity), pp. 275–291. Available from: https://doi.org/10.1108/S2398-601820200000006019.

Madden, R. (2010). *Being ethnographic: a guide to the theory and practice of ethnography*. London: SAGE Publications. Available from: http://ebookcentral.proquest.com/lib/otago/detail.action?docID=743685 [accessed 28 September 2022].

Malinowski, B. (1922). *Argonauts of the western Pacific an account of native enterprise and adventure in the archipelagoes of Melanesian New Guinea*. Long Grove: Waveland Press, Inc.

Marcus, G.E. (1995). Ethnography in/of the world system: the emergence of multi-sited ethnography. *Annual Review of Anthropology*, 24(1), pp. 95–117. https://doi.org/10.1146/annurev.an.24.100195.000523.

Markowitz, F. and Ashkenazi, M. (1999). *Sex, sexuality, and the anthropologist*. Champaign: University of Illinois Press.

Massey, D. (2003). Imagining the field. In: Pryke, M., Rose, G. and Whatmore, S., eds. London: Sage Publications. Available from: http://www.sagepub.co.uk/booksProdDesc.nav?prodId=Book226267 [accessed 6 August 2024].

Miller, D. (2018). Digital anthropology. In: *Cambridge encyclopedia of anthropology* [Preprint]. Available from: https://www.anthroencyclopedia.com/entry/digital-antthropology [accessed 26 December 2023].

Pink, S., Mackley, K.L. and Moroşanu, R. (2015). Hanging out at home: laundry as a thread and texture of everyday life. *International Journal of Cultural Studies*, 18(2), pp. 209–224. Available from: https://doi.org/10.1177/1367877913508461.

Pink, S. and Morgan, J. (2013). Short-term ethnography: intense routes to knowing. *Symbolic Interaction*, 36(3), pp. 351–361. Available from: https://doi.org/10.1002/symb.66.

Qamar, A.H. (2020). At-home ethnography: a native researcher's fieldwork reflections. *Qualitative Research Journal*, 21(1), pp. 51–64. Available from: https://doi.org/10.1108/QRJ-03-2020-0019.

Rivera-González, J., Trivedi, J., Marino, E.K. and Dietrich, A. (2022). Imagining an ethnographic otherwise during a pandemic. *Human Organization*, 81(3), pp. 291–300.

Robben, A.C.G.M. and Sluka, J.A. (2012). *Ethnographic fieldwork: an anthropological reader*. 2nd ed. Chichester: Wiley-Blackwell (Blackwell anthologies in social and cultural anthropology, 9).

Smith, C.W. (2020). I try to keep quiet but my ancestors don't let me. In: George, L., Tauri, J. and Te Ata o Tu MacDonald, L., eds. *Indigenous research ethics: claiming research sovereignty beyond deficit and the colonial legacy*. Bingley: Emerald Publishing Limited (Advances in Research Ethics and Integrity), pp. 127–140. Available from: https://doi.org/10.1108/S2398-601820200000006009.

Sterling, M. (2010). *Babylon East: performing Dancehall, Roots Reggae, and Rastafari in Japan*. Durham: Duke University Press. Available from: https://doi.org/10.1215/9780822392736.

Tsing, A.L. (2011). Friction: an ethnography of global connection. In: *Friction*. Princeton: Princeton University Press. Available from: https://doi.org/10.1515/9781400830596.

Tusing, C. (2023). The fieldwork of never alone: reframing access as relationships of care. In: *The entanglements of ethnographic fieldwork in a violent world*. London: Routledge.

Udupa, S. (2015). *Making news in global India: media, publics, politics*. New York: Cambridge University Press.

Wang, D. and Liu, S. (2021). Doing ethnography on social media: a methodological reflection on the study of online groups in China. *Qualitative Inquiry*, 27(8–9), pp. 977–987. Available from: https://doi.org/10.1177/10778004211014610.

Wolcott, H.F. (2008). *Ethnography: a way of seeing*. California: AltaMira Press. Available from: http://ebookcentral.proquest.com/lib/otago/detail.action?docID=1343764 [accessed 28 September 2022].

3

NAVIGATING RESEARCH RELATIONSHIPS

Ethnography is, at its core, a relational practice: a strategy for developing meaningful knowledge about the social world by connecting with the people in it. Relationships with participants can determine whether you can access the field, how rich your data will be, and whether the process feels good and safe to all involved, as well as what impact the research will eventually have. "What is ultimately at stake for these people?" Lowe et al. recommend asking, and "On whose terms will the research be carried out?" (Lowe et al. 2020, p. 279). In other words, ethnographers must apply an ethical and **reflexive** approach that recognises the axis of politics and power shaping the process while also recognising the human level on which interactions are played out. Thinking about both aspects, this chapter discusses the various roles an ethnographer might take on within their fieldsites and what responsibilities these roles might establish. It covers some of the ways that interpersonal relationships can become central to the process, recognising that ethnographers are often dealing with multiple (and sometimes competing) forms of relationality. This connects onwards to a discussion of wider models of research that define not just relationships between individuals, but between communities and research institutions. As such, it grapples with worries and critiques about extractive research, continuing the important discussion about **decolonising** ethnography through considering what participatory, collaborative, or activist-based ethnographic research might look like and achieve.

DOI: 10.4324/9781003404880-3

PARTICIPANTS BY ANY OTHER NAME

For a long time, the prevailing image of the ethnographer was a romanticised one, centred on a lone figure out in the field (see Chapter 2). But ethnographic data is rarely generated alone. Rather, ethnographic knowledge is always relational, the product of multiple cross-cutting conversations across diverse contexts (Wardle and Gay y Blasco 2011). In other words, ethnographic knowledge is collaborative: generated *with* as well as *about* people.

The words that ethnographers use for the people they engage with in the field will vary. Common ones include '**participants**', '**interlocutors**', '**informants**', and/or '**subjects**'. I primarily use the term 'participant' throughout this book, and the previous chapter discussed a bit about how to identify who might count as a 'participant' in the eyes of an ethics board. But the discussion about how to name and identify participants is bigger than that. Each of the terms mentioned has been in and out of fashion for specific reasons, tapping into both **epistemological** frameworks and ethical values about how knowledge is (or should be) generated. *Who has agency? Who is an expert? Who can speak?* At its very best, ethnography has the potential to collapse the traditional distance between researcher and researched, reconfiguring power relations between the 'subject' and the 'object'. But this does not happen automatically. Rather, ethnographers have to be honest and deliberate in thinking about these dynamics in their own study, not just in terms of the words they will use, but how the voice, expertise, and agency of different people involved can be recognised in the process.

ETHICS IN ACTION, ETHICS IN RELATIONSHIP

Ethnographic research deals with real people and their real lives. The researcher's actions both in and beyond the field can have social, financial, political, or legal ramifications for their participants and for others too. All researchers will need to engage rigorously with institutional ethics protocols, before starting, in order to lay out a plan that fulfils basic criteria for participant safety and informed consent. However, on its own, this usually isn't enough to be able to discern how to act sensitively and ethically within the actual real-world settings in which you work, especially over long periods of time, where you are not only enacting your own plans but responding to others' actions and to emergent situations. Ethnographers can benefit from grounding their own actions in core ethical principles, such as those of nonmaleficence, beneficence, autonomy, and justice (Murphy and Dingwall 2001). In commonsense terms, these make a great deal of sense. We don't wish our actions to harm our participants (nonmaleficence). We hope our research will bring good to our participants and our fieldsites, either directly or indirectly (beneficence). We aim for those we engage with to be able to exercise choice and self-determination in how they engage with us (autonomy). We hope our research doesn't contribute to wider systems or structures that are unfair or oppressive, or that it can actively work against those that do (justice). But deeper and ongoing reflexive work is needed to consider how to apply these principles in the particular social and political settings you are working in. Furthermore, in the actual flux and flow of the field, many ethnographers turn to a framework of relational ethics to govern their choices about how to engage with participants. This approach means that what is ethical and 'good' is worked out in the context of human relationships and governed by *relational* principles such as dignity, respect, and care.

BUILDING RELATIONSHIPS: ACCESS, REJECTION, AND RECIPROCITY

Being an ethnographer involves a commitment to "begin with people" and to centre "the human relationships around which ethnography ultimately revolves" (Campbell and Lassiter 2014, p. 4). The process of building these relationships can be pleasurable, enriching, and deeply meaningful – one of the very best parts of the ethnographic method. It can also be complicated, confusing, and stressful. This is especially true because building trust and rapport with particular people or groups can be essential to gaining access to the field, but can also take time and involve lots of room for misunderstandings, especially if the researcher is moving into new spaces or communities. Having a lead-in period ahead of formally starting fieldwork can be helpful in enabling you to introduce yourself and get to know the people you hope to work with, without being 'on the clock'. This can be approached through the notion of "pre-ethics" (Lowe et al. 2020, p. 279), an idea arising from Indigenous scholars that asks non-Indigenous scholars in particular to take time to learn how their work might actually serve the communities they are studying. They elaborate pre-ethics as a time to:

> situate oneself in a relational field of accountability and collaboration from the very start of a research project to be willing to have one's project radically reshaped by others and to abandon ideas imagined from afar if they prove wrong-headed or otherwise inappropriate by potential collaborators.
>
> (Lowe et al. 2020, p. 279)

This process often decenters the researcher themselves while asking them to remain honest and self-aware about their own agendas in trying to access those spaces and relationships. Key to the process is not *assuming* the project will go ahead.

STORY BOX: Listening to sound and people

Sebastian Lowe has an enduring interest in 'sound worlds'. As an anthropologist and a musician, he has had a particular interest in ngā taonga pūoro – the traditional musical instruments of Māori, the Indigenous peoples of Aotearoa New Zealand. Yet he himself, born in Aotearoa, is Pākehā – that is, someone of New Zealander European, or settler, decent. Lowe was based at an Australian university when he was awarded funding towards his doctorate. He had already completed an MSc on this topic, and on paper, he was ready to start work. Yet a number of people advised him to hold off a little. On their suggestion, he instead travelled back home to Aotearoa New Zealand for five months of 'pre-ethics'. This period of time focused on face-to-face encounters and "deeper deep listening", with the goal to "find ways to meaningfully collaborate with Māori" before officially beginning his research. While the process of pre-ethics was very much about building relationships with potential collaborators – and seeing where he could contribute to what was already happening in the community in ways that were supported by members of the community – for Lowe, it also created a "new relationship with myself" as he thought through who and what he represented, what his responsibilities were, what was at stake in the research, and to whom he would be accountable.

Figure 3.1 Alistair Fraser, a practitioner of taonga pūoro (Māori Indigenous musical instruments), conducting a field recording in Mason Bay, Aotearoa New Zealand, in 2011. Anthropologist Sebastian Lowe approached the album Fraser produced as an 'ethnographic case study', and the two published about it together (Lowe and Fraser 2018).

At any stage in the research project, individuals or whole communities might decide they do not want to work with the ethnographer. This is sometimes termed 'ethnographic rejection' or 'participant refusal'. While this can be an expression of agency on the part of participants (McGranahan 2016), that does not mean it won't be practically disruptive or personally upsetting for the ethnographer. At the same time, it can be an opportunity to learn both about your own research practices, and about the dynamics and sensitivities of your research field.

STORY BOX: When participants google you

Emily Burns had been working for many months to set up a new ethno-graphic research project. Her plan was to study home birth practices among members of a new religious movement in Australia. Towards this goal, Burns had been visiting the community to get to know them, sometimes bringing her own family with her. She was "devastated" to be told suddenly that they had decided not to participate. When she tried to unpack what had happened, Burns learned that although she had already shared her formal participant information documents with them, members of the group had later also googled her and become concerned with what they found, including the project title listed on the university website. In a highly networked age, she reflected later, we have to assume that a lot of 'backstage' or institutional info will be accessible to participants. Once she realised that the rejection wasn't about a lack of rapport, she was also able to see it as useful example of "the non-data of research practice" (Burns 2015, p. 203), meaning refusal could itself reveal something about the social field. In this case it highlighted the complex social position of the group amidst the politically charged topic of home birth in Australia, which left them feeling protective around having their views written about.

Even when general permissions have been given to start a proj-ect, navigating ethnographic research may involve dealing with **gatekeepers**. Gatekeepers are people who have the ability to deter-mine whether we will be allowed 'in' to certain social spaces. This might be about practical sorts of connections (i.e. who will forward you information about meetings, pass on contacts lists, and set up digital permissions), or it might be about social status (i.e. who might others look to to get a sense of whether it is safe, appropriate, or cool to talk to you). At both levels, it is clear that access is not a one-time achievement but rather a series of accesses to different spaces, people, or aspects of our fieldsites. This gives access a "multifaceted, nonlinear, opportunistic, and sometimes serendipi-tous" quality (Cunliffe and Alcadipani 2016, p. 536), which in turn means it relies both on thoughtful pre-planning and on investing enough time to build connections and take up opportunities when they do emerge.

RECIPROCITY AND REMUNERATION

Relationships are typically give-and-take, and this is no less true of fieldwork relationships. There is, however, an asymmetry to the relationship between researcher and participant – not necessarily because of their social **positionality** but because they are each coming to the encounter for different reasons and with different things at stake. Because of this, relationships with participants are not necessarily about *equal* exchange but about working out what it is appropriate to offer of yourself as part of building and honouring the relationship.

> **STORY BOX: When a watermelon ruined an interview**
>
> American anthropologist Chris Boehm was trying to study mental health categories among the Navajo (First Nations) people in the 1960s. At that time, he describes, white people were lumped together in one of two categories: missionaries or foreign spies. To help gain access to people in the community, he worked with a young Navajo man, Howard McKinley Jr, as an interpreter. While Boehm tried generally to follow his interpreter's social cues, he still managed to blunder plenty of times. On one occasion, McKinley was able to arrange an interview with an uncle at the uncle's house. After sitting down on the porch and talking for a while but "not yet getting down to business", Boehm suddenly remembered he had brought watermelon as a gift and went to fetch it from the truck. He was then stunned to find himself sent away before any actual conversation had begun. McKinley had to explain to him, after they had left, that gifting the watermelon after the interview and as a "sign of gratitude and appreciation" would have been appropriate, but that the uncle had taken his choice to present at it at the start as a "power move" – a pushy way to barter for the interview – which he thus promptly rejected (Davis and Konner 2011, p. 95).

Paying participants for their time or assistance can *sometimes* answer *some* of the issues with extractive research (as discussed further on). However, it can also risk turning your relationship into something transactional. The nuances will be very situationally specific. For this reason, it can be controversial, and ethics boards have varying rules and preferences about whether

Figure 3.2 A day-long event for youth workers, organised and funded by Susan Wardell (the author), providing an opportunity for her to talk with more people, run focus groups, and distribute surveys, but also facilitating a rare opportunity for them to gather, network, share a meal, and listen to speakers.

Source: Kampala Uganda, January 2013.

participants should be formally remunerated for their help, lest this be seen as coercive. There are, of course, ways of remunerating people that aren't strictly monetary. Vouchers, personal gifts, or exchanges of time, labour, expertise might be appropriate, again with the need to consider cultural, community, or organisational norms. Indeed, while remuneration is an important ethical question for some people, *reciprocity* is a broader principle that often takes indirect forms, sometimes extended over a substantial period of time. It can also be less about repaying things within individual relationships and more about contributing at the wider level of the community, aligning with ethical principles of beneficence by thinking broadly about how good might be brought back to your participants through your presence. Research designed collaboratively with communities (as the chapter returns to later) has this more overtly built in.

ROLES AND RESPONSIBILITIES IN THE FIELD

Much of the time, ethnographers aren't able to fully anticipate the nature of the relationships they are going to form until they are in the field. Despite this, preparing for fieldwork should involve thinking through what roles you might take up, as these will directly shape the types of relationships you can form. *Are your interactions with others going to be based on some kind of formally designated role? Are you relying mostly on informal social relationships? If so, what kind?* With either type of approach, there can be tensions and complexities: *What does professionalism look like? What is appropriate, ethical, and safe, for all involved?*

BEING EMBEDDED

Not many people hear the term 'research' and picture the sorts of things we do as ethnographers, especially in terms of participant-observation (see Chapter 4). It can be hard to explain ourselves in a way that helps people in the field understand what we want and need from them and what they can safely expect from us in return. Are we a researcher or a guest? A family member or a visitor? A volunteer or an observer? While the answer might ideally be 'both', this can understandably be confusing for individuals or communities who are trying to work out how to interact with us. In addition, people tend to interpret the researcher within whatever social categories they have on hand, which may or may not be accurate, or helpful.

Having a formal role in the fieldsite can help create a reason for your presence and a way to get involved. Ethnographers may be able to gain access as employees, volunteers, teachers, students, and so on. In certain settings, host communities or organisations may value having an ethnographer on board because of some of the skills they bring. Because of this, being 'embedded' can help an ethnographer to establish a mutually beneficial relationship with their collaborators.

STORY BOX: Where there is smoke, there is an ethnographer

Susan Jane Lewis and A.J. Russell are researchers who spent three years embedded with Fresh Smoke Free North East – the first dedicated office for tobacco control in the UK. This involved doing participant-observation at the team's base, attending meetings, shadowing team members, and interviewing staff and members of the advisory panel. They also analysed "the complex web of documents and electronic communications that the team generated" (Lewis and Russell 2011, p. 404). This put them in a position to experience the practices and 'worldview' of the organization, its members, and its partners. But just as important was the fact that they were able to "withdraw, reflect, and work with a critical distance" to analyse this (412). They made use of their **liminality** to give insights back to the organisation so it could make use of them. For example, when they interviewed some of the local coordinators who were members of 'alliances' that the team worked with, they learned of some of the anxieties and challenges these people faced. With permission from interviewees, they gave a confidential report back to the organisation, detailing the worries anonymously, and the core team was then able to reformulate plans for relationships with these alliances with a greater sensitivity to the pressures faced by local coordinators.

Having a formal role can also create social and ethical complexities. It might open up opportunities for certain types of dialogues or interactions, but it will also likely close possibilities for others. Furthermore, because the ethnographer is then occupying two roles at once, it's easy for tensions to arise between the different expectations or obligations associated with each role. Researchers will need to constantly reflect on who they are beholden to, and what ethical boundaries apply to their interactions, including how information they gather is circulated or what it is used for.

In addition, not all roles are formally declared. Ethnographers often take on unspoken or informal social roles in the communities they are in – for example, as a peer, an apprentice, or a family member.

BEING A BEGINNER

The role of the 'apprentice' – where the ethnographer is positioned as a novice who can be guided by an expert in the development of specialised skills or knowledge – has historically been particularly valuable for ethnographers (Downey et al. 2015). This is sometimes a formal role, but not always. In fact, it links to a broader orientation which is useful for ethnographers – that of 'not knowing'. Positioning yourself as a beginner can be a strategic position because it encourages people in the field to show you things and tell you things. It allows you to ask 'stupid' questions in order to have people explain things that matter to them in their own words (see also Chapter 6). It allows you to make mistakes.

Mistakes in the field can take many forms. They may result from misunderstanding, misjudgement, or miscommunication. They can have a variety of ramifications. It is easy to make social faux pas; to accidentally offend someone or get them offside; to break organisational or social protocol. This can feel horrible, embarrassing, and vulnerable. Taking up a humble stance from the beginning can assist with some of these types of things, allowing both you and your participants to address things as needed and move on. Of course, more serious ethical breaches will need to be addressed in a more formal manner with your researcher supervisors, managers, and/or research ethics board.

BEING POSITIONED

An ethnographer's way of entering their field and forming social relationships will be shaped by their social positionality. This includes gender, race, and class, as well as religion, **culture**, sexuality, education level, life stage, health status, and so on (see also Chapter 4). What matters is how this positions them in relation to their participants, and their participants to them. One key aspect is how these factors might place them as either 'insiders' or 'outsiders' (or a mixture of both) in their particular social fields, and what the implications of this might be.

Being an insider to the field you are studying can sometimes set others more at ease. It can provide practical benefits, such as starting with a baseline familiarity of the histories, social norms, and lingo of that setting, which can help smooth how you move through it and interact with people. This positionality may have deeper benefits, too, in terms of accessing not only social spaces but subjectivities and thus generating different types of knowledge and insight (see Chapter 4). But being an insider can be complex too. For a start, it means the ethnographer is already positioned in the social field through existing social relationships and connections. This will shape how people see them and who will want to talk to them about which topics. It might mean they have pre-established personal, moral, or political commitments, which may complement *or* pull against their goals as an ethnographer.

What is also extremely important to realise is that the designation of 'insider' is not straightforward. In fact, insider status is typically partial, defined by layered and multifaceted identities. As Allen-Collinson (2013, p. 287) notes,

> it is debatable whether anyone can ever be deemed a complete member of any culture, subculture or social group . . . rather, it might be more accurate to posit that there are degrees of insiderness and outsiderness, which change over time, place and social context.

In the same vein, Lila Abu-Lughod (1996) explored the complex experiences of the 'halfie' ethnographer, who may be assumed to have access to their community and yet not be entirely accepted. Even just deciding to start research in your own community (i.e. putting on the ethnographer's 'hat') can add a layer of outsiderness or social distance, which is why the idea that ethnographers often work from or within '**liminal**' spaces can apply to even insider ethnographers. What is a key takeaway, overall, is that "the ethnographer's identity as a researcher is not fixed in the way typical of other forms of research" (Bell 2019). Because of this, they often experience multipositionality.

BEING POWERFUL

> **STORY BOX: Schoolyard conundrums**
>
> It took Jocelyn D. Avery fourteen months to get the approvals she needed to conduct fieldwork at a school for children with severe intellectual disabilities, in Australia (Avery 2019). This included approvals from her research institutions, from the research site, and from the State. After all of this work, Avery was able to conduct two semesters of participant-observation at the school, utilising a degree of insider knowledge as the parent of a child with 'special needs' herself. But even with all the formal boxes ticked, she experienced unforeseen "ethical conundrums". Confidentiality was not always straightforward. For example, she was asked by a senior staff member to recount events she had witnessed during lunch recess when another staff member had interacted with students in a way that was being questioned by the school's managers. Who did she have a moral responsibility to if she felt the student's point of view had not been upheld, but did not want to get offside with the teachers either? Avery found there could be tensions between privacy and welfare when observing potential issues with a student's health or home life, both of which were well beyond the scope of her own research. Avery describes her "cross-cutting" subjectivities as a student, researcher, friend, parent, staff member, volunteer, and noneducator as both enhancing her experience as an ethnographer and providing challenges in how she would adhere to ethical principles.

Ethnography can be very well suited to investigating the experiences of marginalised, stigmatised, or vulnerable people. Ethnographers can provide detailed, contextualised data that is based on real-world observations, and built over time through relationships of trust. However, there are typically considered to be higher risks of harm in working in such areas, and institutional ethics protocols often ask for extra controls for research on topics considered 'sensitive' or people considered 'vulnerable'. Vulnerability is not always about individual factors, either, but can be about structural and systemic factors. This means that communities as well as individuals can be vulnerable. It also means there can be complex and intersecting factors affecting degrees of agency, power, and vulnerability *within* any given field. This is another reason for reflecting on your positionality in relation to your participants. This is necessary to be able to act ethically and safely, with recognition of unseen factors that may shape how free they feel to say no, to express themselves in particular ways, or to push back or correct you.

However, while most discussions of research ethics identify the ethnographer's position as one of power over their participants, this is not always true. Asymmetry in research relationships can go both ways. Projects in which an ethnographer is intentionally studying people who hold higher power or status than they do have been described as 'studying up'. Gaining access to the sorts of sites where powerful people live out their everyday or social lives, in order to conduct fieldwork there, can be particularly challenging.

STORY BOX: Being the dumb model

Graduate student Ashley Mears was sitting at a café when a modelling recruiter approached her. Little did they know Mears had already given up a part-time modelling career some years ago. But she turned this unexpected encounter into an opportunity. On signing up with a new agency Mears began a covert ethnography, planning to explore the construction of desirability and value in the fashion modelling market. Doing this as a model was intended to help her look at the 'backstage' of a glamorous industry in order to "see the structure of the world I was studying" (Mears 2013, p. 31). But although she had become an insider in one sense, Mears also realised she was 'studying up' from a precarious and low-status position. While she worked hard to establish rapport with fellow models as well as with bookers and clients, being a model meant she found herself "confined to a narrow bandwidth of behaviour" (26). It was almost impossible, for example, to ask people about the treatment of women in the fashion industry. On the flipside, she found it a useful strategy to lean into 'playing dumb' in order to ask people to explain different terms or particular aspects of the industry.

A related method is that of 'elite interviewing'. This is where a researcher seeks to set up more focused conversations with people who hold more knowledge or power than them, which might include professionals, policy-makers, and other political, corporate, or cultural leaders. Even where these people are not the main focus of the study, they may provide a different sort of information towards the ethnographer's goal of a holistic understanding of their field. Again, it is important to remember that the category of the 'elite' can be variable, dynamic, and contextual (Mbohou and Tomkinson 2022).

Ultimately, ethnographic fieldwork is always a two-way encounter between two (or more) active agents, each with their own interests, vulnerabilities, and agendas at play. The axes of power can shift in complex ways. Individuals, organisations, and communities can all have their own reasons for engaging with a researcher. This could be as simple (and significant) as wanting to feel heard. It could be about prestige by association. It could be that they have a certain hope of what you can offer them, which, in turn, may or may not be accurate. *Will you bring money with you? Will your research raise their profile and bring more funding/tourists/exposure?* The immersive nature of ethnographic research can mean a researcher is easily drawn into political manoeuvring, where they themselves become either a commodity or a liability (Cunliffe and Alcadipani 2016, p. 550). At the same time, the researcher must continue to reflect honestly on what their own goals or agendas are – *Are they invested in building a career? Getting a larger grant? Publishing a prominent book? Pleasing a client? Fulfilling a political goal?* – considering how their pursuit of these goals is affecting their participants and their work.

BEING TRANSPARENT

Informed consent is a cornerstone of contemporary research ethics. It requires those an ethnographer interacts with in the field to be clearly informed about their purpose there. This can be a difficulty with ethnography in general. *How do you announce your presence or role in a larger group setting? Who, specifically, has to consent*? Some types of ethnography make this trickier still – for example, digital ethnography, where the type of infrastructures that facilitate social interaction blur the boundaries between covert and overt research, and the issue of the 'lurking' ethnographer has become one of the most recurrent challenges. In some unique circumstances, ethnographers have deliberately employed a technique of covert ethnography to provide insights into elite or closed worlds. This is controversial, as deliberate deception is strongly discouraged by contemporary ethics boards and by most professional associations, as something that violates the (unknowing) subject's privacy and autonomy. While many ethnographers have been critical of this practice, David Calvey has argued that covert ethnography does have a legitimate place for some areas of research, for example, when looking at criminality, deviance, or illicit subcultures (2013).

These types of studies often highlight other conundrums, including that ethnographers may, at times, learn things or witness events that discredit or threaten their participants in some ways. This might involve illegal, unethical, or violent activities. In some studies, this may, in fact, be part of what you are trying to document. You then face decisions about if, when, and how to share this information with relevant authorities. The obligation to report wrongdoing or harm may tug against other obligations – to keep confidentiality, for example. Ethnographers will need to consider what the impact on their participants will be if they do report things. *What harm will be rendered? Alternatively, what harm will be rendered by not doing so? And what will happen to your relationship with that individual or that community based on this decision?*

As a side note, it is important to make the distinction that being transparent about your role as a researcher does not have to mean disclosing *every* aspect of your personal identity, especially where this may not only threaten the success of research relationships but endanger the safety of the researcher. For example, a researcher who is queer, working in a setting where there is discrimination against queer people, may understandably choose not to disclose this (Lewin and Leap 1996).

BEING COMMITTED

Specific people who become particularly valuable sources of insight in the field can be described as **key informants**. These in particular can be relationships that are sustained over the long term. Still, most ethnographic projects end, or at least the formal periods of field-work do. That does not mean the *relationships* (with individuals, communities, or organisations) end. In fact, being "in it for the long haul" is in fact a hallmark of ethnographic research, as Kristy Nabhan-Warren suggests: "We show up, we hang around, we ask questions, we listen, we learn from dialectical engagement, and we return" (2022, p. 27). Researchers who are not normally based in the fields they have been researching need to think carefully about how to exit the field when the time comes (Franco and Yang 2021). They will also have to think about returning, with the awareness that this doesn't have to be about conducting further research or even about presenting research findings to the community but can simply be about honouring existing relationships.

INTERPERSONAL RELATIONSHIPS

The long-term, immersive nature of much ethnographic research and the intense experience of listening to people's stories or witnessing their lives can create feelings of closeness. Ethnographers may form interpersonal relationships of a variety of different types in the field. The quality of these can also shift over time. In other words, they can take on an "undesigned" quality (Bell 2019), sometimes becoming as much personal as they are professional. In turn, these personal relationships can become a significant part of the research process and experience, not only offering sources of rich information, but creating the potential for collaborative knowledge-making and long-term reciprocity, through establishing ethical commitments.

STORY BOX: The PhD and the peddler

At twenty-seven years old, after finishing her PhD, Cuban-American anthropologist Ruth Behar moved to the town of Mexquitic, in Mexico, to try and decide what to do with her life (Behar 2015). Shortly after this, in 1983, she met Ezperanza, an illiterate fifty-three-year-old farmer and street peddler. Others in the town warned her about Esperanza, who "was known to be fiery, rude, and a witch". Despite this, a unique friendship emerged, and ten years later it yielded a text that would become one of Behar's most famous: *Translated Woman* (1993). Behar describes herself and Esperanza as "co-mothers" of this book, which presented "The story of her life [. . .] entangled with the story of how we came to know each other and why I was the one who wrote it down". It was published first in English, then Spanish, and while Esperenza couldn't read it herself, her grandchildren could. Behar visited many times over the years, even after she moved back to the USA. In 2015, marking the occasion of Esperanza's death at age eighty-four, Behar described this book as "a souvenir of a time when two women, two strangers, dared to embark upon a conversation with each other".

INTIMACY AS METHOD

While not everyone forms a genuine affection for their participants, many do, and sometimes friendships can emerge quite organically. Empathetic or affective relationship with participants can help to create the type of rapport that can lead to unique insights, to the point that friendship is sometimes seen as a research method in and of itself. Friendships can help further dissolve traditional distinctions between informant and ethnographer, facilitating various forms of collaboration. As Banerji and Distante put it, intimacy can "promote an atmosphere of trust and exchange that, in turn, enables the creation of knowledge on a truly inter-subjective and dialogic basis" (Banerji and Distante 2009, p. 38).

Figure 3.3 Ethnographer Anurima Banerji and fellow dancers/friends/artistic collaborators prepare for a performance of Odissi Indian classical dance as part of an 'intimate ethnography' that focused on subverting the gaze upon the racialised body.

Source: Photo credit: Ilaria Distante.

Depending on the setting – and perhaps particularly common for those who live with their participants – kin-like or 'adopted kin' relationships can also be established over time. These, too, can provide a degree of recognition or insider status in a field. At the same time, these relationships position the researcher within an existing set of relations in the field and create their own sets of social obligations. Furthermore, some ethnographers are not forming relationships from scratch but working within existing relationships, for example, where they are interviewing immediate friends or family members. This can be part of an autoethnographic approach, an insider ethnography, or what Waterston and Rylko-Bower have called an "intimate ethnography" (Waterston 2019). In all cases there can be complexities to interpersonal relationships. Many of these complexities relate to the potential asymmetries of power or diverging goals in or for the relationship. Someone may be simultaneously a friend and a participant. They may be both a family member and the subject of written publications. One form or relationality may predate the other or may change over time. One may have potential advantages or disadvantages for the other. In fact, dealing with multiple and intersecting forms of relationality is often a major part of the experience of ethnographic fieldwork.

It is also true than an ethnographer won't *always* like their participants (Kleinman and Copp 1993). It is possible to be angry, frustrated, disappointed, or upset about specific interactions or situations. At times, this can extend to a relationship that becomes fraught or antagonistic overall. Keeping an eye on the (physical and emotional) safety of yourself as a researcher, as well of that of the participants, is essential (see also Chapter 2). Try to notice any patterns or changes over time, including any escalation of behaviours that feel uncomfortable or unsafe, or a change in the nature of the relationships. Important to note also is that although a variety of different forms of interpersonal relationship may be woven usefully into **fieldwork**, ethical boundaries prevent romantic and sexual relationships being deliberately pursued as part of this. Not all sexual behaviours are welcome or invited, either, so anything approaching this type of relationship needs to be handled with great care and through transparent conversation with supervisors, managers, and so on to ensure that everyone is safe.

DEALING WITH DIFFERENCE

Ethnographic research can sometimes require an ethnographer to engage with people across lines of cultural, religious, or political difference. Even if this is their intent, ethnographers may find it challenging to build relationships with people whose actions, beliefs, or identities they find morally discomforting or even repugnant (as I expand in Chapter 4). Common values that ethnographers apply to research relationships – including collaboration, reciprocity, solidarity, and advocacy – may be called into question when the people that the ethnographer is studying are associated with oppressive, violent, or discriminatory regimes or groups, for example. *Is there a limit to the type of difference that fieldwork relationships should encompass? Or a particularly safe or ethical way to encompass it?*

> **STORY BOX: Drinking with white nationalists**
>
> When he first began studying radical white nationalist groups in Nordic countries in 2010, Benjamin R. Teitelbaum intended to engage "as a neutral, dispassionate observer". As a professor in a college of music in Colorado coming into a new social field, his initial plans were to conduct formal interviews and to monitor public gatherings from a distance. This became both unsustainable and unideal, especially as he saw "their participation and engagement increased as they sensed a new willingness on my part to understand their lives" (Teitelbaum 2019, p. 419). He began to shift his approach so that in time, as well as observing public demonstrations and attending concerts, Teitelbaum "laughed, drank, dined, and lived with them" (414). His strategy became to "cultivate close long-term relationships with nationalists fed by honesty, personal exchange, and trust" (414). This has gained him a mixture of praise and critique from all sides. While he acknowledges that leaning into the ethnographic paradigm of scholar-informant solidarity with these groups created a "morally volatile" situation, he defends it as an approach leading to the type of ethnographic knowledge unlikely to be gained through other forms of research.

The role of the ethnographer in taking up a moral, ethical, or political stance in relation to their **fieldsites** has been much debated. Since its early days, ethnographic research has aimed to apply a lens

Figure 3.4 Anti-nuclear protestors outside of the prime minister's office in Tokyo, Japan, marking long-lasting tensions in the domain of nuclear politics, which the ethnographer, Makoto Takashi, faced when he undertook research via embedding himself in a policy organisation.

Source: Image credit: Makoto Takahashi.

of **cultural relativism**. This is the idea that the cultural practices and norms being studied should not be compared back to the researcher's own (personal, institutional, or cultural) norms, but understood within their own context. This helps avoid ethnocentric (and historically Eurocentric) bias. This, however, is not the same thing as *moral* relativism, which suggests there are (or should be) no universal ethical standards to judge human practices by. The work of people like 'militant anthropologist' Nancy Scheper-Hughes (1995) has encouraged ethnographers to refuse a traditional scientific detachment and, instead, take up a moral position, pick a side, and be positioned. This is often essential for conducting activist ethnography. However, others have critiqued the way that establishing a particular moral standpoint can close off possibilities of understanding other people or groups involved in the same setting. And of course, there is the question of *whose side* they should take up, given history is also peppered with examples of ethnographers creating deeply questionable alliances with political or military projects, creating a need to distinguish between 'collaboration' and 'collusion' (Trundle 2018).

MODELS OF RESEARCH: INSTITUTIONS, POLITICS, AND PARTICIPATION

Ethnographic research is reliant not only on interpersonal relationships but on the relationship between research institutions and communities. Many of the topics already covered in this chapter – for example, around access, reciprocity, positionality, and power – apply at both levels. In other words, "the ethics and politics of ethnography are not clearly separable" (Murphy and Dingwall 2001, p. 339).

CRITIQUES OF EXTRACTIVE RESEARCH

Ethnographers have been haunted by worries about the potentially extractive nature of research for many years now. Research is extractive if it gains knowledge from a community and uses this in settings outside of the community, for someone else's benefit (e.g. for the benefit of a government body, commercial entity, or organisation, or for the career advancement or financial gain of the individual researcher). This has been a broad critique of Western research paradigms. Ethnography has certainly not been exempt.

Ethnographic research can be particularly challenging to navigate in settings where communities have been subject to unequal power dynamics in their relationship with researchers in the past. This has often been the case in the relationships between research institutions and Indigenous communities, where knowledge has been used in decontextualised and reductive ways, or operationalised against them to justify the exercise of power or control. It also applies to many other marginalised groups. This can understandably create distrust towards researchers along with additional layers of vulnerability. Contemporary ethnographers need to be especially mindful in such settings. The writing on decolonising research addresses ways forward from this directly. It acknowledges too that there is no 'one size fits all' solution, since it is not just about an individual researcher's intentions, but also how they are positioned in relation to their field. As Lily George et al. express,

A researcher should ask themselves "Who am I? What right do I have to be doing this research?" (George 2018). A researcher must recognise how their inclusion in particular socio-historical and cultural contexts influences the way they see, judge or interpret other people's worlds as well as their own.

(George et al. 2020, p. 7)

Research done by people from indigenous or other minority communities with their own people is often considered the safest and most appropriate. But there are still many pathways for an 'outsider' researcher to work towards doing research in these spaces, or with these communities, in sensitive and ethical ways. This is usually built on relationships and on listening, as the earlier discussion of pre-ethics highlighted. Ultimately, all ethnographers need to take up accountability in the way they design and enact their research, including around issues such as ownership and sovereignty of data, participation and involvement of community members, the use and benefit of outputs, and representational ethics. This requires them to take up reflexive work at quite a personal level, but it may also mean reflecting on the limits provided by the systems and structures they are part of.

PARTICIPATORY AND COLLABORATIVE APPROACHES

Many authors have argued that that the best response to extractive Western research paradigms is to work actively to establish "a 'participatory consciousness', which sees the redistribution of power over 'the research questions, research paradigms, project design and methodology, control of the project and ownership of the data collected'" (Bishop and Glynn 1999, in George et al. 2020, p. 284). I have already established the idea that in a deeper sense, all ethnographic knowledge is relational, but there are many ways to shift more deliberately into collaborative or participatory processes. One of the frameworks most widely recognised across the social sciences is participatory action research (PAR), a **qualitative** method, pioneered in the late 1940s, that focuses on researchers and participants collaborating to generate knowledge intended to create social change. But as discussed throughout other chapters of this book, there can be a variety of different approaches and frameworks used towards these goals. Luke Eric Lassiter's work on **collaborative** ethnography is taken as a touchstone by many people, offering a both theoretical and methodological approach that emphasises deliberate efforts to work with participants, letting their goals, choices, and insights shape the project at *every* point in the process (Lassiter 2005).

The overarching challenge is that that 'participation' can mean different things to different people in different contexts. Collaborative approaches that aim to build meaningful engagements can run up against all sort of relational, practical, and everyday complexities, sometimes failing, becoming impractical, or generating "ethical conundrums and moral dilemmas" of their own (Trundle 2018, p. 89). This type of work often requires great care and patience, and ongoing negotiation, based in reflexivity, humility, and openness to opportunities to learn or readjust practices throughout the process.

LOCALISED AND INDIGENOUS MODELS

Ethnographers often bring well-intentioned ethical codes with them into the field. But it can help to think about the frameworks or approaches that will be recognised or valued by your collaborators, not just by your own institution. Indeed, many Indigenous communities have their own well-elaborated frameworks for ethical relationships or partnerships. Often, these "are local, particular and do not aspire to become 'the' way of doing things", so they can be embraced as existing in multiplicity (George et al. 2020, p. 3). While some Indigenous research frameworks are only appropriate for those who are themselves Indigenous to use, others provide sets of principles that can also be applied by non-Indigenous researchers as a guide for relating to those communities. For example, in my own setting of Aotearoa New Zealand, 'Kaupapa Māori' research describes an approach grounded in a Māori **worldview** and a commitment to addressing historical inequalities. While non-Māori researchers cannot claim to 'do' Kaupapa research, they can learn key principles from it that will help them to engage in healthy, productive, and culturally safe partnerships with Māori (Tuhiwai Smith 2012).

STORY BOX: Learning at her mother's table

Long before her research started, Tarapuhi Vaeau (nee Bryers-Brown) benefitted from "the indigenous, female, and experiential knowledge pass down to me through the many women that have sat at my mother's kitchen table" (Vaeau 2015). It was also her parents who provided the call to do research with her own people. Responding to this call, Vaeau's master's research project involved seven months of fieldwork, guided by a 'Kaupapa Māori' framework, which focuses on knowledge as contextual and relational. Her 'field' was her whanau (extended family) home in Whanganui, where she used participant-observation, interviewing, and **autoethnographic** techniques to explore her family's everyday life and engagements with health organisations in order to understand more about structural violence and historical trauma as part of Māori experience. "Being an insider researcher involved being intimately immersed, and not separate from, the emotionally rich worlds and experiences of my participants" she explains (18). Part of her work involved participant-observation while living in the community full-time with her whanau, at which times she often wrestled with how to signal to her participants that she was putting on her 'ethnographer's hat' and to navigate keeping her identity as a group member while making sure her participants knew when they were being observed. This approach nonetheless allowed her to focusing on building on existing trust and commitments and "prioritising the knowledge her whanau and community have developed about themselves" (17), rather than outsider knowledge, thus fulfilling a core part of the Kaupapa Māori approach.

MINERS OR MOLES?

Critical, reflexive, and **postmodern** turns in the social sciences were all valuable opportunities for ethnographers to confront the wider structural contexts in which their work was undertaken. In some spheres, these processes generated a degree of cynicism about ethnography. *Was it really possible for ethnography to get outside its colonial, imperial, power-laden origins? How could ethnographers truly relate ethically to those they studied?* Other voices, resurging in the 1990s, pushed back against what had become described by some as a trend of negativity or navel-gazing with reminders that "Seeing, listening, touching, recording can be, if done with care and sensitivity, acts of solidarity" (Scheper-Hughes 1995, p. 418, in Robben and Sluka 2012, p. 24). These voices were part of a 'compassionate turn' and involved in a call for 'engaged fieldwork'. They highlighted the fact that ethnographers share not only the mundane daily worlds and activities of their research participants but also a "subjective space" in which they become implicated in each other's lives, suggesting this provided ideal opportunities for witnessing and thus for both ethical and political engagement (Robben and Sluka 2012, p. 25).

STORY BOX: Documenting the undocumented

Amidst the many forms of violence against undocumented people and their families, it seemed "profoundly immoral" to academics Carolina Alonso Bejarano and Daniel M. Goldstein to simply write about the topic from afar (Bejarano et al. 2019, p. 9). Instead, as they started an ethnographic project on undocumented people in New Jersey in 2011, they felt "compelled" to work with a community organisation that was fighting for immigrant rights. Two years later, Mirian A. Mijangos García and Lucia López Juárez – both undocumented immigrants who had never heard of ethnography before starting this work – joined the project as research assistants. Over time, they evolved into collaborators and "full ethnographers". "The results were transformative [. . .] By the end of our project, ethnography had changed them and they had changed ethnography" (13). As they continued their work, ethnography and activism became "indistinguishable", with the rich, detailed data emerging proving "a much more robust source for academic analysis and writing than ordinary field-work methods would have provided", and their academic learnings at the same time making them more effective activists. The team linked their approach to a decolonising ethos, explaining that "We didn't simply want to extract data, but to use what we learned to throw a monkey wrench into the workings of both the U.S. deportation regime and the academic-capitalist machinery of scientific research" (10).

Returning to the queries about 'extractive' research, feminist STS scholar Juno Salazar Parreñas presents an interesting metaphor. She suggests that while ethnographers do 'extract' knowledge, they can

Figure 3.5 Mirian and Carolina perform a scene from their play *Undocumented, Unafraid*, developed in collaboration with a variety of people, and aiming to present some of their findings about immigrant workers rights in an accessible form (2015).

Source: Illustration by Peter Quach

perhaps aim to be less like multinational mining companies and more like moles: "maintaining a sensitivity to where they tread and where they might dig" and letting the desire to go beneath the surface be part of "making new connections by digging channels in the dark" (2023, p. 454). In such a way, ethnography makes use of its fundamental lens of situated, collaborative, and reflexive knowledge-making to become a tool for both witnessing and connection. Taken up by a range of people from any number of backgrounds, it can facilitate both personal transformation and public advocacy, helping to de-centre knowledge-making practices and establish research relationships that hold value for everyone involved.

CHAPTER SUMMARY: KEY POINTS

- The success of ethnographic research depends on the ability of the researcher to build relationships with individuals and communities. The best, most ethical, and most effective way of doing this will be very much specific to the context, and will require ongoing reflexivity and accountability.
- Interpersonal relationships in the field can be a pleasure and a challenge. They provide a rich basis for genuinely collaborative work and establish commitments and responsibilities.
- Researchers may take up a variety of different (formal and informal) roles in their field sites. Their social positionality will also shape their access and relationships, based in part on degrees of insiderness or outsiderness which are typically partial and shifting anyway.
- Contemporary ethnographers have been haunted by worries about the extractive nature of research. Critical and reflexive attention to the power relationships between ethnographers and participants is needed, but in doing so – and in pursuing participatory and collaborative approaches – ethnography retains the potential for sensitive, meaningful, and politically engaged work.

RECOMMENDED FOR FURTHER READING

Davis, S.H. and Konner, M. (2011). *Being there: learning to live cross-culturally*. Cambridge: Harvard University Press.

George, L., Tauri, J. and MacDonald, L.T.A.O.T. (2020). *Indigenous research ethics: claiming research sovereignty beyond deficit and the colonial legacy*. Bingley: Emerald Publishing Limited.

Lassiter, L.E. (2005). *The Chicago guide to collaborative ethnography, Chicago guides to writing, editing, and publishing*. Chicago: University of Chicago Press.

McGranahan, C. (2016). Theorizing refusal: an introduction. *Cultural Anthropology*, 31, pp. 319–325.

Murphy, E. and Dingwall, R. (2001). The ethics of ethnography. In: *Handbook of ethnography*. London: SAGE Publications Ltd. Available from: https://doi.org/10.4135/9781848608337

Smith, L.T. (2013). *Decolonizing methodologies: research and indigenous peoples*. 2nd ed. London: Zed Books.
Trundle, C. (2018). Uncomfortable collaborations: the ethics of complicity, collusion, and detachment in ethnographic fieldwork. *Collaborative Anthropologies*, 11(1), pp. 89–110.

REFERENCES

Abu-Lughod, L. (1996). Writing against culture. In: Fox, R.G., ed. *Recapturing anthropology: working in the present*. Santa Fe: School of American Research Press, pp. 137–162.
Allen-Collinson, J. (2013). Autoethnography as the engagement of self/other, self/culture, self/politics, selves/futures. In: Holman Jones, S., Ellis, C. and Adams, T.E., eds. *Handbook of autoethnography*. Walnut Creek, CA: Left Coast Press.
Avery, Jocelyn D. (December 2019). Ethical dilemmas and moral conundrums: negotiating the unforeseen challenges of ethnographic fieldwork. *Anthropology in Action,* 26(3) pp. 1–9. https://doi.org/10.3167/aia.2019.260301.
Banerji, A. and Distante, I. (2009). An intimate ethnography. *Women & Performance: A Journal of Feminist Theory*, 19(1), pp. 35–60. Available from: https://doi.org/10.1080/07407700802655547.
Behar, R. (1993). *Translated woman: crossing the border with Esperanza's story*. Boston: Beacon Press.
Behar, R. (2015). Goodbye, comadre. *Beacon Broadside: A Project of Beacon Press* (blog). Available from: https://www.beaconbroadside.com/broadside/2015/01/goodbye-comadre.html.
Bejarano, C.A., Juárez, L.L., García, M.A.M. and Goldstein, D.M. (2019). Introduction. In: *Decolonizing ethnography*. Durham: Duke University Press (Undocumented immigrants and new directions in Social Science), pp. 1–16. Available from: https://doi.org/10.2307/j.ctv11smmv5.6.
Bell, K. (2019). The "problem" of undesigned relationality: ethnographic fieldwork, dual roles and research ethics. *Ethnography*, 20(1), pp. 8–26.
Burns, E. (2015). "Thanks, but no thanks": ethnographic fieldwork and the experience of rejection from a new religious movement. *Fieldwork in Religion*, 10(2), pp. 190–208. Available from: https://doi.org/10.1558/firn.v10i2.27236.
Calvey, D. (2013). Covert ethnography in criminology: a submerged yet creative tradition. *Current Issues in Criminal Justice* [Preprint]. Available from: https://www.tandfonline.com/doi/abs/10.1080/10345329.2013.12035980 [accessed 7 August 2024].
Campbell, E. and Lassiter, L.E. (2014). *Doing ethnography today: theories, methods, exercises*. 1st ed. Chichester, West Sussex; Malen, MA: Wiley-Blackwell.

Cunliffe, A.L. and Alcadipani, R. (2016). The politics of access in fieldwork: immersion, backstage dramas, and deception. *Organizational Research Methods*, 19(4), pp. 535–561. Available from: https://doi.org/10.1177/1094428116639134.

Davis, S.H. and Konner, M. (2011). *Being there: learning to live cross-culturally*. Cambridge: Harvard University Press. Available from: http://ebookcentral.proquest.com/lib/otago/detail.action?docID=3301024 [accessed 5 December 2023].

Downey, G., Dalidowicz, M. and Mason, P.H. (2015). Apprenticeship as method: embodied learning in ethnographic practice. *Qualitative Research*, 15(2), pp. 183–200. Available from: https://doi.org/10.1177/1468794114543400.

Franco, P. and Yang, Y. (2021). Exiting fieldwork "with grace": reflections on the unintended consequences of participant observation and researcher-participant relationships. *Qualitative Market Research: An International Journal*, 24(3), pp. 358–374. Available from: https://doi.org/10.1108/QMR-07-2020-0094.

George, L., Tauri, J. and MacDonald, L.T.A.O.T. (2020). *Indigenous research ethics: claiming research sovereignty beyond deficit and the colonial legacy*. Bingley: Emerald Publishing Limited. Available from: http://ebookcentral.proquest.com/lib/otago/detail.action?docID=6370619 [accessed 28 December 2023].

Kleinman, S. and Copp, M.A. (1993). *Emotions and fieldwork*. London: SAGE Publications. Available from: https://doi.org/10.4135/9781412984041.

Lassiter, L.E. (2005). *The Chicago guide to collaborative ethnography*. Chicago: University of Chicago Press (Chicago guides to writing, editing, and publishing).

Lewin, E. and Leap, W. (1996). *Out in the field: reflections of lesbian and gay anthropologists*. Champaign: University of Illinois Press.

Lewis, S. and Russell, A. (2011). Being embedded: a way forward for ethnographic research. *Ethnography*, 12(3), pp. 398–416.

Lowe, S. and Fraser, A. (2018). Connecting with innerlandscapes: Taonga pūoro, musical improvisation and exploring acoustic Aotearoa/New Zealand. *Journal of New Zealand & Pacific Studies*, 6(1), pp. 5–20. Available from: https://doi.org/10.1386/nzps.6.1.5_1.

Lowe, S.J., George, L. and Deger, J. (2020). A deeper deep listening: doing pre-ethics fieldwork in Aotearoa New Zealand. In: George, L., Tauri, J. and Te Ata o Tu MacDonald, L., eds. *Indigenous research ethics: claiming research sovereignty beyond deficit and the colonial legacy*. Bingley: Emerald Publishing Limited (Advances in Research Ethics and Integrity), pp. 275–291. Available from: https://doi.org/10.1108/S2398-601820200000006019.

McGranahan, C. (2016). Theorizing refusal: an introduction. *Cultural Anthropology*, 31(3), pp. 319–325.

Mears, A. (2013). Ethnography as precarious work. *The Sociological Quarterly*, 54(1), pp. 20–34. Available from: https://doi.org/10.1111/tsq.12005.

Murphy, E. and Dingwall, R. (2001). The ethics of ethnography. In: *Handbook of ethnography*. London: SAGE Publications. Available from: https://doi.org/10.4135/9781848608337.

Nabhan-Warren, K. (2022). Participant observation: embodied insights, challenges, best practices and looking to the future. *Fieldwork in Religion*, 17(1), pp. 26–36. Available from: https://doi.org/10.1558/firn.22582.

Ntienjom Mbohou, L.F. and Tomkinson, S. (2022). Rethinking elite interviews through moments of discomfort: the role of information and power. *International Journal of Qualitative Methods*, 21. Available from: https://doi.org/10.1177/16094069221095312.

Parreñas, J.S. (2023). Ethnography after anthropology. *American Ethnologist*, 50(3), pp. 453–461. Available from: https://doi.org/10.1111/amet.13201.

Robben, A.C.G.M. and Sluka, J.A. (2012). *Ethnographic fieldwork: an anthropological reader*. 2nd ed. Chichester: Wiley-Blackwell (Blackwell anthologies in social and cultural anthropology, 9).

Scheper-Hughes, N. (1995). The primacy of the ethical: propositions for a militant anthropology. *Current Anthropology*, 36(3), pp. 409–440.

Teitelbaum, B.R. (2019). Collaborating with the radical right: scholar-informant solidarity and the case for an immoral anthropology. *Current Anthropology*, 60(3), pp. 414–435. Available from: https://doi.org/10.1086/703199.

Trundle, C. (2018). Uncomfortable collaborations: the ethics of complicity, collusion, and detachment in ethnographic fieldwork. *Collaborative Anthropologies*, 11(1), pp. 89–110.

Tuhiwai Smith, P.L. (2012). *Decolonizing methodologies: research and indigenous peoples*. London: Zed Books. Available from: http://ebookcentral.proquest.com/lib/otago/detail.action?docID=1426837 [accessed 25 May 2020].

Vaeau, T. (2015). *"He reached across the river and healed the generations of hara": structural violence, historical trauma, and healing among contemporary Whanganui Māori*. Thesis. Open Access Te Herenga Waka-Victoria University of Wellington. Available from: https://doi.org/10.26686/wgtn.17013635.v1.

Wardle, H. and Gay y Blasco, P. (2011). Ethnography and an ethnography in the human conversation. *Anthropologica*, 53(1), pp. 117–127.

Waterston, A. (2019). Intimate ethnography and the anthropological imagination. *American Ethnologist*, 46(1), pp. 7–19. Available from: https://doi.org/10.1111/amet.12730.

4

'BEING THERE' AS A PARTICIPANT-OBSERVER

Many types of researchers use instruments to gather data. Ethnographers don't typically have any kind of specialised tools or equipment. Instead, they have themselves: their minds, their bodies, and their social senses. These are all put to use through one of the distinctive practices of ethnography, **participant-observation**. Participant-observation involves being present, looking for opportunities, following leads, listening, and learning as you go. It can't be fully planned out ahead of time, which can be exciting and also vulnerable. With this in mind, rather than offering a strict 'how to', this chapter starts by addressing the principle of immersion and discussing some of the different possibilities for (and barriers to) participation in **the field**. It then unfolds into ideas around how an ethnographer might fine-tune themselves as an instrument of data gathering, arranged into sections on mindwork, bodywork, and heartwork. These sections cover topics such as attention, defamiliarisation, bracketing, sensory attunement, detachment, social participation, emotion work, and empathy. Each makes clear the need for **reflexivity** around the scope or limits of knowledge you might access through participant-observation but also highlights the rich potential for gaining vivid, detailed, and holistic understandings of social worlds when the whole self becomes a means of learning.

'BEING THERE' AND BEING IMMERSED

Ethnographic knowledge is largely premised on the idea that 'being there' makes a difference. Participant-observation can be seen as the cornerstone of this: a particular *way* of being there through

DOI: 10.4324/9781003404880-4

being immersed *and* being involved. But immersion is multifaceted concept, involving practical immersion, intellectual immersion, *and* emotional immersion (Dumont 2023, p. 441). Flexibility can be needed to work out what immersion might look in different *types* of fields – for example, when studying a geographically centred community versus an online fandom, a closed institution versus a global political movement, a community festival versus a certain type of professional. Sociologist Erving Goffman (1989, p. 125) argued that

> [T]he standard technique is to try to subject yourself, hopefully, to their life circumstances, which means that although, in fact, you can leave at any time, you act as if you can't and you try to accept all of the desirable and undesirable things that are a feature of their life. . . . [Y]ou are in a position to note their gestural, visual, bodily response to what's going on around them and you're empathetic enough – because you've been taking the same crap they've been taking – to sense what it is that they're responding to.

But while for some ethnographers, this might be best achieved by living in a place or with a group, in other cases, that might be impossible or unnecessary, and immersion might instead come through strategies such as taking up a particular role in the field (see also Chapter 3).

DOING PARTICIPANT-OBSERVATION

Many research methods involve some form of observation. Often, these focus on observing human behaviour in a controlled or simulated environment, for example, a research lab. Ethnography, however, focuses on **naturalistic** research, which means observing people in real-world settings. In addition, and unlike **positivist** tropes that value a neutral, distanced observer, the ethnographer is acknowledged (and encouraged) to get involved in the social world they are studying: to live, work, play, and experience alongside the people they are trying to understand. The goal is to be in a position to not just observe, but interpret the *significance* of what is observed, by building up an understanding of the social context from the

Figure 4.1 A Canadian researcher (part of a multidisciplinary team) and a Quechua farmer working together to remove kernels of corn from dried cobs, in Rio Negro, Argentina; participant-observation providing a chance to be part of daily life, build rapport, be useful to participants, and ask questions while doing so.

Source: Photo credit: Jenny Cockburn (2004).

inside. Because of this, an ethnographer is not just an observer of **participants** but a *participant-observer*.

Ethnographers need informed consent to conduct research this way; that is, they need others that they are sharing the space with to be aware of who they are and why they are there and to be okay with it (see also Chapters 2 and 3). But the settings in which ethnographers may want to do participant observation are diverse, and being a participant-observer may encompass varying *degrees* but also varying *types* of participation, as the next few sections unfold. Depending on how the individuals and communities we are working with understand the term 'researcher', we might confuse our participants greatly when we come along and just want to . . . hang out? . . . help out? . . . tag along and take notes?

Participating in everyday life and in social life

While ethnographic **fieldwork** can sometimes involve participating in activities with a special or formal quality – for example, cultural performances, religious rituals, or important life events – this isn't a necessity. In fact, the things ethnographers participate in most often and which often generate the most insight are the mundane and everyday parts of people's lives. The **ethnographic lens** of **holism** suggests that all parts of social life are connected, meaning it is possible to learn a lot about a **culture**, society, or organisation from observing everyday details. Pink (2015) notes that the near-universal quality of some activities – such as talking, sitting, dressing, eating, or walking – makes them particularly interesting things to participate in and analyse because the differences between how different social groups undertake them can be compared. It is worth noting, though, that, it can be *harder* to gain access to some parts of everyday life, and especially those that happen in private

Figure 4.2 A painting by ethnographer Priyanka Borpujari from her study of aging and cell phone use in India, based on a photo taken in her fieldsite, showing her Maa in her bedroom trying to nap after lunch but also enjoying using her cell phone.

spaces, than it is to attend events or performances that are done in or for the public.

Ethnographers often want to participate not only in specific activities but in everyday social life – in the in-between moments that stitch life together. In fact, participant-observation is sometimes jokingly described as a method of 'deep hanging out'. But the ethnographer's ability to participate socially can depend on acceptance and access granted by participants, which depends, in turn, on how participants 'read' them. Impression management often has to begin early in the process, for example, when you are first introducing yourself to potential participants. But while some factors may be malleable – for example, you can adjust your behaviour or modify your normal styles of self-presentation – others are tied to fixed qualities which we cannot change or hide yet which will shape our ability to participate socially in certain spaces. These include things like body size or type, skin colour, gendered or sexual features, dis/ability, and so on. Any of these can shape how participants perceive you, and therefore how (and when and where) they are willing to include you.

STORY BOX: Being half Muslim and almost female

Hélène Thibault is a political scientist who studied conservative Muslim communities in Tajikistan from 2010 to 2011. But she is not herself Muslim. As a white scholar and an agnostic, she felt that she was "clearly an outsider" (2021, p. 403). Yet because she expressed an interest in Islam as a focus of her research and chose to wear a headscarf out of respect, her participants began to describe her as "half-muslim" (403). Thibault was aware that her foreignness, gender, and marital status all shaped the way she was able to participate in different parts of the social field. While as a single woman, she was at risk of being seen as 'shirking' her gendered responsibilities to marry and bear children, she felt that her foreignness protected her from outright critique about this. In fact, as some other female ethnographers have also experienced, she found she was treated as something of a 'third gender' and was able to move in both male spaces and female spaces – for example, sitting at the men's table, where they were offered alcohol. While this was useful for her access to male participants, there were also risks of local women begrudging her the 'special treatment' and thus reducing her rapport with them, or access to their social circles.

Participating in specialised activities, participating in subjectivities

Sometimes, the people an ethnographer is studying are undertaking highly skilled or professionalised activities, for example, athletes, performers, clinical professionals, religious practitioners, and so on. Because of this, unless they are already an insider with specialised training, ethnographers may be able to do participant-observation in the wider social setting, but with barriers to participating in particular activities. Some ethnographers have committed an enormous amount of time and effort as an 'apprentice' in order to learn specialised skill sets. For other researchers, observing, talking, listening, and asking people about things they can't participate in directly (see Chapter 6) is an equally effective way forward.

STORY BOX: Danger, dirt, and robots

In an era of 'Industry 4.0', robotics are being introduced to many industrial settings to automate production and enhance productivity. Ned Barker and Carey Jewitt were more than ready to provide a human perspective on this. To do so, they undertook an ethnographic project in two industrial settings – a glass factory and a waste management centre – where this change was happening. Using a multisited approach, the researchers spent one week in each of these settings, shadowing the shifts of manual labourers. They attended training, worked alongside the workers and robots, and shared break time and conversation, aiming for something of a "tactile apprenticeship" (Barker and Jewitt 2022, p. 106). A sensory ethnography approach encouraged them to explore aspects of the work that were 'dirty' or 'dangerous', with a particular interest in touch and how this was being reshaped by AI and collaborative robots. But there were limits to the forms of dangerous work they could engage in themselves, in a hands-on way. In fact, they were told, as part of the parameters of their access, that "you cannot touch anything to do with the machinery". To fill these gaps, they asked the trained machine operators to demonstrate, used video and photos to get 'guided walk-throughs', and conducted semi-structured interviews so they could listen to the workers talk about their experiences in detail.

Issues with participant-observation can also occur when the types of activities the ethnographer wants to study are unsafe, unethical, or illegal. Given the complexities of many fieldwork

Figure 4.3 Left: A swabbing brush/stick used to apply oil (dobe) to the molds in the machinery that can be seen in the background of this photo. Right: Field researcher learning how to swab using touch controllers in virtual reality.

settings, ethnographers need to be able to "adjust their levels of involvement and participation" to fit the circumstances as things develop (Li 2008, p. 111) as well as thinking actively about alternative ways to gain insider understandings of practices or experiences they can't participate in directly.

A **subjectivity** is, put simply, someone's way of being in the world. This includes their way of sensing, perceiving, thinking, and being in their body, which, in turn, is shaped by biology, culture, gender, political alignments, economic position, and more. Ethnographers are often interested in understanding the subjectivities of their participants. But at the same time, they are already positioned subjects themselves, with their own ways of being in the world. It can be important for ethnographers to distinguish between activities they can participate in and subjectivities that they may not be able to occupy even if they *are* participating temporarily in the same activities. Researchers who are able to conduct insider ethnography or **autoethnography** contribute significantly by producing knowledge that might otherwise be precluded by some of these issues. But as the next sections will discuss, there are other ways ethnographers might try to approach an insider experience while not overclaiming the extent of their knowledge.

STORY BOX: A hopepunk with front-row tickets

Donna Passero Crilly is a former media professional whose Masters research focused on 'hopepunk', a countercultural artistic moment arising from the era of Donald Trump's first presidential term. She explains that "As an American with a front row ticket to the crises and the subsequent rise of hopepunk, I was just as much 'in it' as I was watching it" (Crilly 2023, p. xii). For this reason, her approach was *autoethnographic*, combining personal reflections with a survey and what she called "cyber archaeology" – an analysis of a range of digital texts and media artefacts. Crilly writes that

> I draw from personal experiences because of my inability to fully separate myself from hopepunk's context. Likewise, I found myself participating in movements of resistance, protest, and art creation. I marched in protest of inequities and injustices toward Black people. I danced in a punk music video created for the purposes of activism and art.
>
> (xvi)

Over a period of two years, she also produced her own Tumblr blog, reblogging hopepunk texts, and keeping a journal of **fieldnotes** and memos, thoughts, observations, anecdotes, and content related to hopepunk, meaning that her processes of participation and recording overlapped. When all this was subjected to analysis, **coding**, and **triangulation**, Crilly found that it offered a way to understand the topic through "epiphanies", that is, with a focus on life-altering moments in her own experience that offered insight into both the hopepunk movement and the social and political contexts that birthed it.

MINDWORK

Anthropologist Harry Wolcott (2008) wrote about ethnography as form of mindwork: a way of *seeing* as well as a way of *looking*. In this section, I discuss the practice of participant-observation by extending Wolcott's idea into my own framework of mindwork, bodywork, and heartwork. Importantly, in practice, ethnographers are whole people, and these three aspects are very much entangled. They are also not *comprehensive* categories. Rather, they are intended to serve here as a way to structure discussion about some of the lived complexities of immersion, engagement, and presence in the field as a participant-observer.

THE PRACTICALITIES AND POLITICS OF ATTENTION

A participant-observer is continually making decisions about who and what to focus on in their field: what to participate in, what to observe. But there is a politics to this. Ethnographers need to be continually reflexive about who and what is included or excluded as part of their field of attention. This requires being reflexive about what they might be most primed or willing to see, which may involve countering implicit assumptions about what or who is 'important'. This could relate to status, gender, or age (for example, excluding children's experiences) or subjects typically not considered serious enough for academic attention (such as jokes or feelings). Ethnographers gain enormous benefit from noticing the things that others don't.

Even so, there is often a lot going on in a **fieldsite**. The issue of not being able to pay attention to everything at once might be mitigated by conducting ethnography in teams or in pairs. The latter was especially common in earlier eras, when husband-and-wife teams often went to live and work in the field together. This enabled them to participate in and experience different aspects of the field, often based on their different gendered **positionalities**, but also simply based on their ability to notice more between them. More often, though, an ethnographer is solo in the field. The challenge is also deeper than a distribution of attention. It is about the *type* of attention. Ruth Behar describes 'participant-observation' as an oxymoron, a practice that is "split at the root" (1996). French ethnologist Jeanne Favret-Saada similarly writes that balancing between the 'participation' and 'observation' parts of participant-observation is "about as straightforward as eating a burning hot ice cream" (2012, p. 438). She suggests it can generate a form of "split experience" in which the ethnographer has to decide when to give precedence to the part of them that is experiencing and being affected by what is happening and when to focus on "the part that wants to record the experience in order to understand it, and to make it into an object of science" (2012, p. 443). Different forms of recording can help by capturing details in a form that can be revisited later (see Chapter 5), yet technology, too, remains a complex aspect of presence in the field.

DIFFERENCE AND DEFAMILIARISATION

In the earliest eras of ethnography, fieldwork was typically conducted in settings distant from the ethnographer's home (see Chapter 2), with the idea that cultural and geographic differences could help to create the sort of **objectivity** assumed necessary to analyse the social world. This idea has been deconstructed many times over. Indeed, it is probably the minority of contemporary ethnographic projects that involves any sort of cross-cultural experience. Yet even without this, fieldwork may still involve moving into unfamiliar social, bureaucratic, professional, organisational, or linguistic settings. As such, vectors of difference can still play a role in the mindwork of the ethnographer, positioning them to be able to notice taken-for-granted features of the field. As they do, the lens of **cultural relativism** (see Chapter 3) can be essential to avoid '**othering**' people or practices that seem strange just because the researcher is comparing them to norms in their own more familiar settings.

On the other hand, ethnographers who are working in settings more familiar to them – for example, doing 'insider ethnography', 'native ethnography', 'backyard ethnography', or 'ethnography at home' – may face different experiences and different challenges. While there are significant benefits to starting with an insider perspective, there can also be a risk of being "blinded by the familiar" (*Bolak 1996*, p. 109 in Thibault 2021). To counter this, they will need to practice 'making the familiar strange', a process also described as **defamiliarisation**. In many ways, defamiliarization is at the heart of research for *all* ethnographers.

> **STORY BOX: Bracketing the inner swimmer**
>
> Gareth McNarry had been a competitive swimmer and a swimming coach in the UK, before he went on to take up a PhD in a department of sports and nutrition. However, his research turned him back towards a study of competitive swimmers. To do this, McNarry conducted partici-pant-observation with a university swim team, taking up role as a volun-tary assistant coach, focusing on two intense five-week periods of immersion. This role was only possible because of the previous experi-ence that made him a 'cultural insider' to the field. It gave him lots of opportunity to observe the swimmers in and around the pool and be immersed in the embodied and sensuous experience of that setting. However, McNarry also described the need to 'bracket' some of his pre-existing knowledge. For example, when attending a course on coaching methods, he realised they were teaching "something that was so familiar [. . .] that it had been missed from his own reflections" (McNarry et al. 2019, p. 147). This led him to consider other forms of bracketing that could help to "challenge some of his own personal phys-ical-cultural knowledge about and experiences of competitive swim-ming and coaching". Although as both swimmer and coach, he held "the lived experience, swimming sensorium, embodied memories, and tech-nical language" of the field, he also needed to "step aside to some degree from this 'known' lifeworld, in order to make the familiar strange and problematic, and to subject it to scrutiny and questioning" (144).

Defamiliarisation is about seeing things afresh. Trying to defamil-iarise a familiar setting or activity is not the same thing as pursuing objectivity or trying to create emotional distance from the subject. Nor is it about trying to scrub clean the pre-existing understandings you have about the social field. Rather, it is about *becoming aware* of these and able to articulate them, at which point they can become useful information. Writing can be one effective tool for defamiliari-sation. When I teach my first-year students about this, I ask them to do an exercise in which they write a description of a place or event they are extremely familiar with (e.g. a supermarket, a rugby game, a birthday party, a classroom) as if they are an extraterrestrial alien observing it for the first time. Asking questions can be another strong approach. Powell et al. (2022, p. 202) describe their development of walking workshops that focus on a series of verbal prompts to help people with "dislodging the taken-for-granted" facets of walking: *"If we were to start crawling on all fours, what new stories might be produced about the places we walk?" "What are the felt forces of our walk?" "From whom was this path, this land, taken? Who is forced to walk elsewhere?"* Other creative, playful, or experimental activities, such as drawing or mapping exercises, can also assist with noticing, sensing, or thinking about a setting in a different way.

BELIEF AND BRACKETING

Trying not only to understand but to *adhere to* the participants' way of thinking can be part of pursuing immersion (Dumont 2023). Yet beliefs, values, and **worldviews** cannot be put on and off at will. Participating in the activities of a particular community is not the same as participating in their "inner lives" (Engelke 2002), and this is further complicated by the fact the researcher doesn't come in as a blank slate but is already a positioned and socialised subject. **Bracketing** is the practice of gently setting aside previous theories, assumptions, beliefs, or expectations in order to engage more openly with what emerges from the fieldsite. Like parentheses in a sentence, bracketing doesn't mean the things contained within the bracket are gone forever or even that they will be totally excluded from analysis. In fact, this wouldn't be a good idea at all. Nonetheless, the researcher may choose to *temporarily* set some things aside for certain parts of the process (for example, when participating in an activity or listening to a participant speak), then carefully bring them back in while interpreting findings later, as part of reflexive practice.

Sometimes, bracketing is about values or expectations, while at other times, it deals with truth claims – beliefs about the nature of the universe, reality, and so on. Bracketing can help researchers avoid making a judgement about whether something is 'true' long enough to establish a detailed description of how it is understood and experienced by those who *do* believe it. This is valuable for people studying religion, the occult, the paranormal, conspiracy theories, alternative spiritual beliefs, and so on. It is not uncommon for a researcher to be grounded in different **ontological** frameworks than the people they are studying. While in the past, the researcher's job would be seen as being to provide an objective, rational, scientific angle, this is not always the ideal for ethnographers, who want to explore, in a nuanced way, the insider perspective. However, some things are more easily bracketed than others, and an outsider researcher can't simply *choose* to believe.

What they can do is choose to bracket their disbelief or skepticism and, instead, adopt a position of methodological agnosticism. This means that for the purposes of the research, they focus on what they don't know or cannot know for sure, in order to retain openness to the truth claims that their participants present. Indeed, active engagement in the field, as Dumont suggests, involves viewing the beliefs or ideals of research participants as credible at the very least, because they are important for explaining more observable features of the social world, such as why people behave in certain ways (2023, p. 445).

Beyond this, an approach which allows for the possibility that participants' beliefs or worldviews could be 'real' has led to startling experiences for some ethnographers. There are many examples and reports of ethnographers observing happenings that went beyond or challenged their own understandings or those of Western rationality.

STORY BOX: Seeing spirits in the flesh

English-American anthropologist Edie Turner worked alongside her husband, Victor Turner, to conduct participant-observation among the Ndembu people in Zambia in the 1960s and 1970s. They both wrote different accounts of what they had observed and experienced. In one, Edie Turner described observing a ritual focused on drawing out a spirit called the ihamba – described locally as a sort of "spirit tooth" – from a woman called Meru. As she watched, she recounts that "I saw with my own eyes a large thing emerging out of the flesh of her back. This thing was a big gray blob about 6 inches across, a grey opaque plasma-like object" (1999, p. 46). She shared in an interview many years later that many other scholars thought this was "crazy" (Engelke 2000, p. 852). And yet:

> I was certain it happened to me. I didn't actually see a tiny little tooth coming out of the skin. I saw the spirit object, a gray blob, come out. [. . .] And one does not retract things like that, you know? I know it's hard for people, but if they begin to take in a little of the reports they hear [. . .] then we can get somewhere. We haven't sufficiently grappled with these issues, and yet they don't go away. There are always more coming up.
>
> (852)

Both Edie and Victor also wrote about their own experiences with converting to Catholicism later in their lives, reflecting openly on what this meant for their research.

There is a benefit in being open to changing your mind, changing your view, or changing your understanding of the world. Despite this, for a long time, there were criticisms against ethnographers who were seen as sliding too far into the 'insider' subjectivity, something described pejoratively in the past as 'going native'. This concept has been critiqued as being based in a colonial kind of logic that perpetuates assumptions about who can and can't produce valid ethnographic insights. Yet ethnography has come to recognise a variety of positions from which the ethnographer might experience, know, and speak. This includes the perspectives of those who are already insiders to the culture, value system, or belief that forms the subject of research. The idea of the native 'speaking back' (Jacobs-Huey 2002) (through native ethnography, or insider ethnography) lent power to a range of historically othered people, enabling them to reclaim their positioned perspective as a valid basis for knowing. In a similar way, standpoint **theory** provides a framework for recognising research that is openly and purposefully positioned in a particular moral, political, or religious framework as valid in its own right; though scholars have also emphasised that the mainstream, secular academy remains more

Figure 4.4 Susan Wardell (the author) participating in a Christian baptism in Lake Victoria, Uganda, in 2012 as part of her doctoral fieldwork.

receptive to some standpoints than others (Howell 2007). This pairs importantly with the acknowledgement that *all* knowledge is, in fact, positioned, so the goal of ethnography is transparency and reflexivity, rather than a pursuit of the 'myth' of objectivity.

BODYWORK

Doing ethnography is always an embodied experience (Pink 2015). But it was not always acknowledged as such. This idea has had to be established against other (Enlightenment, Western, masculinist) traditions emphasising the ideal of an objective, rational, disembodied scholar. A lot of this work was done by feminist scholars, and has been seen as a vital shift by ethnographers of religion, health, disability, work, sports, dance, and many other subjects. Indeed, even if the topic of the research does not seem to be directly about the body, all of social life is embodied in some way. The body senses, apprehends, learns, and holds knowledge; it is a "mode of cognition" in its own right (Pérez 2011, p. 666). As such, participant-observation asks ethnographers to "build upon their bodily experience as a medium to access a new world of perception" and a valuable route to understanding their participants' ways of thinking and acting in the world (Dumont 2023, p. 446). As sociologist Ashley Mears puts it, "We must tune our bodies into the field so we can feel what our informants feel, to get the logic and sense of their social world right into our bones" (2013, p. 21).

BODIES AND RELATIONALITY

Bodies orient us towards each other and towards the material world. Participant-observation allows ethnographers to learn, in a hands-on way, about how bodies produce relationality. This includes a focus on things such as how people move (kinesthetics), how they touch (haptics), and how they position themselves in space (proxemics). Participating in social life offers opportunities to observe, reflect on, and practice this and more. Of course, the rules and meanings of these practices in any given setting relate not only to 'the body' as a general category but also to *specific* bodies whose gendered, racialised, sexualised, and/or dis/abled qualities may be laden with social meanings that lead to more particular forms of embodied interaction. Ethnographers often need to confront the way that their own specific body is shaping the way they are in relation with other people in their field at an everyday level.

Paying attention to the body is also useful in settings that involve co-presence with larger numbers of people in a particular space – for example, as part of religious rituals, cultural or musical performances, sports events, political rallies, or protest marches. Some of this may point to the role of place and space, and to experiences of *emplacement,* as key to social life. At the same time, features such as music, rhythm, sound, and visual symbolism can have an effect at the level of the body, while also connecting to emotions, beliefs, or meanings. Some ethnographers have drawn on a phenomenological approach to understanding this. Phenomenology focuses on the subjective ways that the senses, emotions, perception, will, desire, **temporality**, memory, and cognition come together to create *experience*. As an ethnographer, you can again attune to this by being present in an embodied way.

As a participant-observer, you may also want to consider the natural and built environments you are participating in, and which your body is also part of orienting you to. *What landscapes, architectures, substances, materials, objects, tools, or technologies are shaping social life there? Are there particular objects that have particular significance or agency in that setting?* This can be explored by observing but also through noticing how *you* engage with the aspects of the material environment. This is no less applicable with digital technologies or in online environments, which are often assumed to involve disembodied forms of sociality but, in fact, involve very particular embodied modes of being and sensing. Operating through material technologies held, worn, and touched, physical activities of looking, tapping, clicking, and scrolling can constitute forms of social (and moral and political) activity online. The question of what participant observation means in regards to virtual environments, and in relation to these sorts of social activities, is challenging. *How can other people's embodied actions be observed? What does it mean to participate?* In this area too, reflections on both material assemblages and embodied experiences as a *participant*-observer can fill some of these gaps, helping ethnographers achieve an approach that, whilst distinct from the forms of participation used in a place-based study, are still "embodied, embedded, and everyday" (Hine 2020).

STORY BOX: Dancing along

Every year, around the festive season, choirs of Bakgatla-ba-Kgafela people gather at open-air performance grounds for a ritual dance and choir competition. This is called 'dikopelo'. Keletso Gaone Setlhabi is a member of one of those choirs who decided to both enjoy her culture and write about it, deliberately blurring her roles as academic and performer. This enabled her fieldwork to focus on the "lived place, lived body, lived time, and lived human relations" of dikopelo (Setlhabi 2024). Setlhabi analysed four performances, one which lasted for seventeen hours. Because she couldn't take notes whilst performing, her memory became her field notebook. She also interviewed other choir members. Her embodied experiences directly shaped her decisions about what to record and how. For example, she explained that it was when she felt the feeling of "newness" in her body after donning a fresh uniform that she felt compelled to capture the field through photographs (see Figure 4.5). This led to a tension between an obligation to record images and worries about missing the performance. But the overall approach allowed her to capture her own vivid experience – from the pride of uniforms to the sound of the whistle call to the mapping of time through the movement of natural elements to the transcendence through music – and to use this to analyse how the cultural meaning of dikopelo is tied to its material, sensory, and embodied qualities.

Figure 4.5 Keletso Gaone Setlhabi pictured near the choir's whistle blower in a performance in 2016.

Source: The photo was taken by an audience member on Setlhabi's request.

THE SENSES, THE SENSORY, AND THE SENSUOUS

Ethnography, in many ways, had a head start in valuing experiential knowledge. Nonetheless, a multidisciplinary 'sensorial turn' in the 1990s re-emphasised how valuable the senses are in understanding and analysing social life. Sensory ethnography, at its most simple, is about "attending to the senses in ethnographic research and representation" (Pink 2015, p. 3). This includes recognising the way the senses link to other aspects of social life, including values and judgements, identities, forms of social organisation, and so on. More radically, it promotes techniques that help the ethnographer in acknowledging and cultivating *their own* capacities as a sensor (Lynch 2022). Lynch suggests that this can represent a shift from more "conventional" methods of participant-observation towards something like "participant sensation" (2022, p. 194).

STORY BOX: Sensuous knowledge of sacrifice

Elizabeth Pérez nearly missed "what was going on right under my nose". As a historian of African Diasporic religions, she was conducting fieldwork on the Lucumí religion. Her plans were to conduct participant-observation in a female-led house of workshop on the South Side of Chicago. Pérez initially assumed this would be about attending ceremonies and then analysing their meaning. The house kitchen, where food and animal materials were prepared for the rituals, was at first just where she "busied" herself, waiting for events to start or finish. However, she soon discovered that this was where young members were mentored and information was transmitted – for example, reciting myths while plucking and cooking to prepare sacrifices for the spirits. As Pérez let herself be put to work in the kitchen, it became the "micro-site" of her research. This wasn't just about listening in; it was also about gaining "somatic knowledge" and becoming "enskilled". As she experienced the visceral, sensuous qualities of the Afro-Cuban religion – the proximity to blood, working with meat, various rich cooking smells, the sounds of goats ready for slaughter – she learned that "the olfactory, auditory, and haptic dimensions of participant observation [can] inform the investigative process, often rather unpredictably" (Pérez 2011, p. 666).

Placing yourself in the everyday social settings of your research participants can provide a way of "experiencing the sensory rhythms and material practices of that environment" (Pink 2015, p. 99). But sensory ethnography needs its own forms of reflexivity. Sensory perception doesn't occur *a priori* culture or socialisation but is always already embedded in categories of cultural meaning and interpretation. Most cultures give primacy to some senses over others. As such, it is helpful for the ethnographer to think about *which* senses they are tuning in to. Some fieldsites or topics might lend themselves more obviously to focusing on a particular sense, while at other times, challenging yourself to consider senses you hadn't previously factored in as important might generate new insights. This includes going beyond the Eurocentric construction of the 'five senses'. Here are some other senses not included in those:

- Pressure
- Itch
- Thirst
- Hunger
- Thermoception (sense of temperature, heat, cold)
- Proprioception (sense of the position and movement of different body parts in relation to each other and the material environment)
- Interoception (sense of awareness of internal bodily functions)
- Equilibrioception (sense of balance)
- Stretch receptors (awareness of particular states/movements in lungs, bladder, stomach, gastrointestinal tract)
- Tension sensors (awareness of particular states in muscles)
- Nociception (sense of pain)
- Chemoreception (sense of chemical substances, including through taste but also through the bloodstream, e.g. responding to carbon dioxide or oxygen levels)
- Magnetoreception
- Time
- Energy

Sarah Pink suggests ethnographers see themselves as taking on the role of a "sensory apprentice" (2015, p. 103), that is as someone who learns to purposefully orient, reorient, or disorient themselves through learning different modes of sensuous perception. *Which senses are most often valued, noticed, discussed, or cultivated in your fieldsite? Which might you learn to cultivate in your own body that you are less familiar with?* Pink is also an advocate for technology as a tool that can help with tuning in to specific sensory aspects of a field. For example, photography may help you attune to visuality or spatiality, and sound recordings may help you listen more deeply or in a different way.

SOCIALISED AND SITUATED BODIES

Bodies are biologically and physiologically diverse from birth. They are also malleable and porous, responding and adapting to environments. People deliberately modify the appearance and functionality of their bodies for a variety of purposes, including through the use of worn or implanted technologies, or ingested or applied substances. Furthermore, bodies are *cultural* objects – socialised into particular ways of moving, sensing, and being in the world based on factors such as culture, class, gender, sexuality, age, health status, and more. It is important for *all* ethnographers to acknowledge the ways that the experiences they have whilst participating might differ in key ways from those of their participants; either in the actual sensations, or in the meaning and significance of that experience. They can also then seek to document the tacit knowledge of participants themselves, using different methods of recording or listening (see Chapters 5 and 6).

> **STORY BOX: Shaking (almost) like a shaman**
>
> American anthropologist Robert Desjarlais studied illness and healing among the Yolmo Sherpa, an ethnically Tibetan people in the Nepal Himalayas, in the late 1980s. During his thirteen months of ethnographic fieldwork, he apprenticed with an elderly shaman called Meme Bombo. Apprenticing meant a significant amount of what he learned "occurred through visceral, sensory means" (Desjarlais 1992). Over time, working alongside Meme, Desjarlais did learn to go into something like a shamanic trance. During this trance, his body would shake in time with Meme playing the drum. However, he wrote later that "it quickly became clear that what I experienced and demonstrated of trance behaviour was far from identical to what my neighbours were familiar with" (30). In fact, the difference between the bouncing movement that Meme's body undertook and the side-to-side "Jell-O" movement of his own body often caused others in the field to laugh. In a similar way, while he experienced visions, it was not in the same form as the shamanic visions of Meme. As such, while he became "partly socialised for Yolmo trance", he was aware his experience "never escaped the prism of my own cultural reality" (33).

Applying this knowledge reflexively helps to problematise the idea of the ethnographer's body as a neutral 'tool', focusing on how they are always already situated and positioned. This applies to all ethnographers but may be more immediately obvious to some. As Hannah Gibson, a disabled ethnographer, reminds us, "I cannot leave my bodily struggles at the door when I enter a room" (2019, p. 75). She charts the way this has created barriers or challenges in conducting fieldwork but has also sometimes provided alternative avenues of empathy or connection. In a similar way, neurodivergent ethnographer Cinzia Greco writes about autism as "another way of embodying the world" involving the potential to "see, hear, and process information in different ways" – meaning neurodivergence can also be seen as an **"epistemological** position" (Greco 2022) that may provide its own distinct accesses both to embodied experience and social or emotional ones.

HEARTWORK

In ethnographic research, the researcher affects the social field and the social field affects the researcher. Set against older templates of the 'objective' observer, the role of affect, feeling, and emotion in fieldwork has nonetheless been a complicated question for participant-observers. The 'emotion turn' emphasised the need for emotional reflexivity on all accounts, and Kleinman and Cobb suggested discarding the question "Did this researcher's feelings affect the study?" and instead asking "*How* did the researcher's emotions play a part in the data collection and analysis of this group or setting?" (Kleinman and Copp 1993). This is not about trying to remove, subdue, or apologise for emotions but rather about working out how they, too, can be useful tools of perception and understanding about social life and "can lead to important insights into certain aspects of the participants' emotional and social worlds" (Ådland et al. 2021, p. 501).

STORY BOX: Encountering emotion in the basement

Anne Kirstine Ådland had worked as an oncology nurse for many years before she started an ethnographic study of nursing homes in Norway. The six-month period she spent observing and participating in the everyday work of nursing home staff led to the realisation that sometimes, "the researcher can become her own informant" (Ådland et al. 2021, p. 497). This came to a head during the moments when a staff member took her down to the basement to show her the 'cold room' where bodies of deceased residents were temporarily stored. Despite her own familiarity with death, she felt a daunting discomfort as they descended the stairs. Then, to make matters worse, her participant realised he had forgotten his keys and left her alone as he went upstairs to fetch them. In her fieldnotes, she wrote a vivid description of her sense of the uncanny in the long, quiet minutes that followed. It was an experience that awakened her "sensibilities and curiosity" about how the staff might experience encounters with the dead body (500). She was initially "almost embarrassed" to share the experience with her research team. Ethnographers tend to underreport emotions since they are seen as irrational, she explains. Yet the research group wrote collectively later that "In this case, we saw the emotional experience of the researcher as a source of knowledge to be explored and potentially supported through the data provided on participants' experience". In fact, Ådland's uncanny experience developed into an important line of analysis around 'liminality' in relation to nursing staff's experiences of death (Ådland et al. 2021).

INVOLVEMENT AND DETACHMENT

Ethnography, from its earliest days, has relied on a researcher moving back and forward between their position as analytical observer and an engaged participant. To express strong emotional commitments was seen by some as comprising this. The social liminality the ethnographer often experiences shapes the role that emotions might play in participant-observation. *How attached or detached should an ethnographer expect to feel from the people, places, and events they participate in? How will their level of emotional investment shape their ability to see, understand, interpret, and analyse?* Favret-Saada suggested that it is both methodologically and theoretically useful for a participant-observer to let themselves be affected: to experience intensities instead of staying emotionally distanced (2012, p. 437). But others have argued there needs to be at least a *part* of the self that is kept separate in order to be analytical, meaning that ethnographers need an ability to balance involvement *and* detachment. This, in many ways, can turn into a value-laden question. Ethnographers are often best to establish a stance based on their chosen frameworks, given that the approach relates closely to feminist traditions, and questions related to moral and political engagement in activist ethnography.

Those working on topics to do with suffering, violence, or trauma may have to think carefully about "a balance between empathy and (emotional) distance" for a different reason (Weiss 2023). It is possible for an ethnographer to experience vicarious traumatisation in witnessing, or hearing about, certain form of violence, discrimination, or suffering as part of their fieldwork. The safety and well-being of the researcher needs to be factored in with care and with appropriate professional support and boundaries as well.

SOCIAL PARTICIPATION, EMOTION WORK

The processes of participant-observation can generate a strange, intense, and sometimes bewildering spectrum of emotions. Instead of ignoring or suppressing these, it is worth paying attention to the specific actions, contexts, and relationships they are tied to. Some of the negative feelings arising from fieldwork will be around particular situations or topics (as I return to shortly). Others may be about the research process itself. Participant observation can generate frustration, boredom, or anxiety. Social fields all have their own rules and norms, which a researcher is not always fluent in and which can take time to learn. In other words, fieldwork has a 'dramaturgical' quality, involving self-conscious social *performances* (Scott et al. 2012). Many ethnographers report feeling anxious, reluctant, or uncertain about engaging with others in the field. This is really quite normal and can relate to the particularities of the field or to the researcher's personality, disposition, or neurology (Scott et al. 2012). While the idealised portrait of the **qualitative** researcher may be someone who is bold or extroverted – a 'people person' – in reality, there are no ideal 'type'. As long as they can bring a degree of openness and humility, each ethnographer finds their own social strategies, and introverted or 'shy' ethnographers may find different but just as valid paths to connection with their participants.

At the same time, emotions are not only individual but are themselves social and cultural artefacts. Like the senses, different emotions are valued or devalued, promoted, or repressed in different social fields, and for different groups of people, according to myriad social factors, including things like gendered identities, professional norms, and cultural or religious values. Looking at how people in your field deliberately shape and work upon their emotions to meet social norms might partly achieved by reflecting on the emotional management you have found *yourself* undertaking in the field, in order to be present and to engage in ways that felt socially acceptable.

MORAL FEELING

Ethnographers may encounter strong feelings about the behaviours, beliefs, or practices of the people whose worlds they are trying to immerse in. Emotions are closely tied to moral beliefs, and so many of these are not just a response to a specific situation but an expression of deeply engrained cultural logics, which can also reveal an unconscious value judgement on the part of the ethnographer. These feelings may be on a fairly minor scale, regarding practices that are unfamiliar or uncomfortable – generating what anthropologist Lila Abu-Lughod described in her own fieldwork as a 'squirm' – or they may be on a much more major scale.

As Chapter 3 also discussed, although ethnographers have a long history of studying marginalised or victimised groups, some also seek to build understandings of the dynamics of power, violence, or hatred through studying powerful people or the perpetrators of violence and abuse. *What does it mean to try and gain an 'insider' perspective on the views and values of a Nazi, or a dictator, or a cult leader? What mental and emotional processes would it involve? What forms of participation?* Time spent establishing rapport and building relationships (see Chapter 3) with people in the context of their everyday lives, efforts to listen deeply and try to understand people whose views you wouldn't normally wish to identify with – perhaps even 'bracketing' your own perspectives in order to do so – may feel morally risky. Questions about if and when this should be attempted are best to focus on both what is valuable and what feels safe for particular ethnographers. It can be helped by recognising that the work of listening, documenting, and describing is not the same as agreeing or *endorsing*.

In many ways the ethnographic methodology provides a meaningful response to the 'existential crises' of being called to be proximate to things that one does not agree or identify with personally, in that "it sanctions respect and enables understanding without demanding full participation" (Abu-Lughod in Davis and Konner 2011, p. 3). In other words, the goal of emotional, mental, or moral participation is not necessary in every project. Indeed, some ethnographers may practice a sort of *emotional* bracketing in order to avoid expressing every negative response they experience internally whilst in the field, as a way to move respectfully and safely in those spaces, while still returning to honour their own responses in more private or backstage settings.

EMPATHY, REFLEXIVITY, AND THE LIMITS OF KNOWING

Some writers argue for ethnography to involve an 'empathetic' engagement with others. Favret-Saada had preferred to clarify that allowing oneself to be affected as a participant-observer is *not* the same thing as empathy (Favret-Saada 2012). While empathy involves trying to access others' experiences at a distance through *imagining* them, participant-observation asks the researcher to actually *experience* them. This doesn't mean identifying completely with the insider point of view or claiming to know *exactly* what others experience, however. Indeed, Clifford Geertz also cautioned against the temptation to claim to speak about others as if from within, labelling this "ethnographic ventriloquism" (Geertz 1988, p. 102). Claims of empathy can run the risk of becoming a projection of others' thoughts and feelings, which can lead to researchers unintentionally 'finalising' their participants viewpoints, rather than maintaining a dialogue with them (Smith et al. 2009). As Perez also poignantly expressed it,

> However much I may have wanted to put myself in others' shoes, I was stuck not only with my own feet, but also with my walk, so to speak, shaped in relation to my cultural center

of gravity and sociopolitical location. It was not an option to borrow another's stance to position myself.

(2011, p. 680)

Nonetheless, with appropriate reflexivity, participant-observation *can* bring an ethnographer closer to understandings of some of the joys, tensions, frustrations, and hopes that permeate the social worlds they are studying, especially when paired with other practices of recording, questioning, and listening (covered in the following two chapters) that can help ethnographers' connect their own experiences of the field with participants' voices and perspectives.

CHAPTER SUMMARY: KEY POINTS

- Ethnographic fieldwork uses participant-observation as a key practice, valuing 'being there' but also 'being involved' and 'being immersed', in order to come closer to an insider's perspective.
- Ethnographers will need to be reflexive about the types of things they can and can't participate in, as well as about how similar their experiences will actually be to those of their participants when they do.
- Participant-observation involves 'mindwork' to manage attention, defamiliarise the taken-for granted, and consider if and how to bracket existing expectations, values, and beliefs.
- Participant-observation involves 'bodywork', drawing on embodied and sensory experience as sites of perception and construction of social life, while recognising that different bodies may be positioned, socialised, and interpreted in different ways.
- Participant-observation involves 'heartwork' to negotiate between emotional involvement and analytical detachment, to learn from your own emotions as social artefacts, and to consider the moral and epistemological implications of empathy.

RECOMMENDED FOR FURTHER READING

Adams, T.E., Jones, S.H. and Ellis, C., eds. (2021). *Handbook of autoethnography*. 2nd ed. New York: Routledge. Available from: https://doi.org/10.4324/9780429431760.

Behar, R. (1996). *The vulnerable observer: anthropology that breaks your heart*. Boston: Beacon Press.

Dumont, G. (2023). Immersion in organizational ethnography: four methodological requirements to immerse oneself in the field. *Organizational Research Methods*, 26, pp. 441–458. Available from: https://doi.org/10.1177/10944281221075365.

Kleinman, S., and Copp, M.A. (1993). *Emotions and fieldwork*. London: SAGE Publications. Available from: https://doi.org/10.4135/9781412984041.

Stoller, P. (2011). *The taste of ethnographic things: the senses in anthropology*. Philadelphia: University of Pennsylvania Press.

Van Roekel, E. (2023). Deep ethnography. *American Ethnologist*, 50, pp. 223–235. Available from: https://doi.org/10.1111/amet.13141.

Weiss, N., Grassiani, E. and Green, L., eds. (2023). *The entanglements of ethnographic fieldwork in a violent world*. London: Routledge. Available from: https://doi.org/10.4324/9781003333418.

REFERENCES

Ådland, A.K., Høyland Lavik, M., Gripsrud, B.H. and Ramvi, E. (2021). Death and liminality: an ethnographic study of nursing home staff's experiences in an encounter with the dead body. *Death Studies*, 45(7), pp. 497–507. Available from: https://doi.org/10.1080/07481187.2019.1648343.

Barker, N. and Jewitt, C. (2022). Filtering touch: an ethnography of dirt, danger, and industrial robots. *Journal of Contemporary Ethnography*, 51(1), pp. 103–130. Available from: https://doi.org/10.1177/08912416211026724.

Behar, R. (1996). *The vulnerable observer: anthropology that breaks your heart*. Boston: Beacon Press.

Crilly, D.P. (2023). *Hopepunk as a process: hope in times of crisis*. The California State University. Available from: https://scholarworks.calstate.edu/concern/theses/pn89df234.

Davis, S.H. and Konner, M. (2011). *Being there: learning to live cross-culturally*. Cambridge: Harvard University Press.

Desjarlais, R.R. (1992). *Body and emotion: the aesthetics of illness and healing in the Nepal Himalayas*. Philadelphia: University of Pennsylvania Press.

Dumont, G. (2023). Immersion in organizational ethnography: four methodological requirements to immerse oneself in the field. *Organizational Research Methods*, 26(3), pp. 441–458. Available from: https://doi.org/10.1177/10944281221075365.

Engelke, M. (2000). An interview with Edith Turner. *Current Anthropology*, 41(5), pp. 843–852. Available from: https://doi.org/10.1086/317412.

Engelke, M. (2002). The problem of belief: Evans–Pritchard and Victor Turner on "the inner life". *Anthropology Today*, 18(6), pp. 3–8. Available from: https://doi.org/10.1111/1467-8322.00146.

Favret-Saada, J. (2012). Being affected. Translated by Mylene Hengen and Matthew Carey. *HAU: Journal of Ethnographic Theory*, 2(1), pp. 435–445.

Geertz, C. (1988). *Works and lives: the anthropologist as author.* Redwood City: Stanford University Press.

Gibson, H. (2019). Living a full life: embodiment, disability, and 'anthropology at home'. *Medicine Anthropology Theory*, 6, 72–78. https://doi.org/10.17157/mat.6.2.690.

Goffman, E. (1989). On fieldwork. *Journal of Contemporary Ethnography*, 18(2), pp. 123–132. Available from: https://doi.org/10.1177/089124189018002001.

Greco, C. (2022). *Divergent ethnography: conducting fieldwork as an autistic anthropologist, society for cultural anthropology.* Available from: https://culanth.org/fieldsights/divergent-ethnography-conducting-fieldwork-as-an-autistic-anthropologist [accessed 6 August 2024].

Hine, C. (2020). *Ethnography for the internet: embedded, embodied and everyday.* London: Routledge. https://doi.org/10.4324/9781003085348.

Howell, B.M. (2007). The repugnant cultural other speaks back Christian identity as ethnographic 'standpoint'. *Anthropological Theory*, 7(4), pp. 371–391. Available from: https://doi.org/10.1177/1463499607083426.

Jacobs-Huey, L. (2002). The natives are gazing and talking back: reviewing the problematics of positionality, voice, and accountability among "native" anthropologists. *American Anthropologist*, 104(3), pp. 791–804. Available from: https://doi.org/10.1525/aa.2002.104.3.791.

Kleinman, S. and Copp, M.A. (1993). *Emotions and fieldwork.* London: SAGE Publications. Available from: https://doi.org/10.4135/9781412984041.

Li, J. (2008). Ethical challenges in participant observation: a reflection on ethnographic fieldwork. *Qualitative Report*, 13(1), pp. 100–115.

Lynch, E.E. (2022). *Locative tourism applications: a sensory ethnography of the augmented city.* London: Routledge. Available from: https://doi.org/10.4324/9781003142386.

McNarry, G., Allen-Collinson, J. and Evans, A.B. (2019). Reflexivity and bracketing in sociological phenomenological research: researching the competitive swimming lifeworld. *Qualitative Research in Sport, Exercise and Health*, 11(1), pp. 138–151. Available from: https://doi.org/10.1080/2159676X.2018.1506498.

Mears, A. (2013). Ethnography as precarious work. *The Sociological Quarterly*, 54(1), pp. 20–34. Available from: https://doi.org/10.1111/tsq.12005.

Pérez, E. (2011). Cooking for the gods: sensuous ethnography, sensory knowledge, and the kitchen in Lucumí tradition. *Religion*, 41(4), pp. 665–683. Available from: https://doi.org/10.1080/0048721X.2011.619585.

Pink, S. (2015). *Doing sensory ethnography.* London, UK: SAGE Publications Ltd. https://doi.org/10.4135/9781473917057.

Powell, K., Altuntas, I. and Bricker, M. (2022). Defamiliarizing a walk. *International Review of Qualitative Research*, 15(2), pp. 199–215. Available from: https://doi.org/10.1177/19408447221090659.

Scott, S., Hinton-Smith, T., Härmä, V. and Broome, K. (2012). The reluctant researcher: shyness in the field. *Qualitative Research*, 12(6), pp. 715–734. Available from: https://doi.org/10.1177/1468794112439015.

Setlhabi, K.G. (2024). Dikopelo ritual and performance: the embodiment of place. *Ethnography*. Available from: https://doi.org/10.1177/14661381241251486.

Smith, B., Collinson, J.A., Phoenix, C., Brown, D. and Sparkes, A. (2009). Dialogue, monologue, and boundary crossing within research encounters: a performative narrative analysis. *International Journal of Sport and Exercise Psychology*, 7(3), pp. 342–358. https://doi.org/10.1080/1612197X.2009.9671914.

Thibault, H. (2021). "Are you married?": gender and faith in political ethnographic research. *Journal of Contemporary Ethnography*, 50(3), pp. 395–416. Available from: https://doi.org/10.1177/0891241620986852.

Weiss, N. (2023). Entanglements of fieldwork: an introduction. In: *The entanglements of ethnographic fieldwork in a violent world*. London: Routledge.

Wolcott, H.F. (2008). *Ethnography: a way of seeing.* 2nd ed. California: AltaMira Press.

5

RECORDING, GATHERING, AND PRODUCING ETHNOGRAPHIC DATA

Paul Rabinow once quipped that everything that he did in **the field** was **fieldwork** (2007). Yet not everything done in the field produces 'data', at least not in any tangible sense. The experiences of ethnographic fieldwork can be rich with insights, but an ethnographer needs ways to record what they observe and experience in some form so it can be revisited later. 'Data' can mean different things in different research fields. Ethnography is a **qualitative** research methodology, meaning the data we seek is descriptive rather than numerical, aiming to provide vivid, detailed, and contextualised representations of our fields. Ethnographers also tend to take a **multimethodological** approach, so data won't come in a single form but a variety of forms, something that can contribute to a more **holistic** record of that community, place, and time. Some common forms of ethnographic data include **fieldnotes**, drawings, photographs, audio or video recordings, interview transcripts, media or archival texts, digital texts, or co-produced creative artifacts. What is important for ethnographers to recognise is that recording is never a neutral or passive act but an active, **interpretive**, and sometimes **collaborative** one. This means that practical decisions about how and when recordings are made (or existing sources gathered) relate to bigger **epistemological** questions and ethical responsibilities. While this chapter can't cover all forms of recording in detail, it will aim to discuss the 'how' and the 'why' of a few ways common ways that ethnographers record, gather, or produce data, with this in mind.

DOI: 10.4324/9781003404880-5

MAKING FIELDNOTES

Some types of scholars – historians, archivists, or some media scholars, for example – focus on analysing existing documents or texts. While ethnographers do analyse existing texts, they also create their own texts. Most famously they create a form of text based on fieldwork and **participant-observation**, and which is broadly labelled 'fieldnotes'. These are often gathered together in a journal or diary of some sort and are sometimes taken as a symbol of the identity of an ethnographer. But 'fieldnotes' is also a vague term that is often thrown about with little clarity as to what exactly they should include or what form they should take. Indeed, they can be a somewhat "bizarre genre" (Lederman 1990, p. 72 in Lönngren 2021), a mixture of firsthand observations, gathered information, vivid description, interpretation and analysis, resulting in something "betwixt and between a personal diary and a scientific document" (Jackson 2010).

Producing fieldnotes can be truly enjoyable: calming, creative, and useful. It can also be laborious, tiring, or boring. There are no strict rules about what sort of volume of fieldnotes you need to

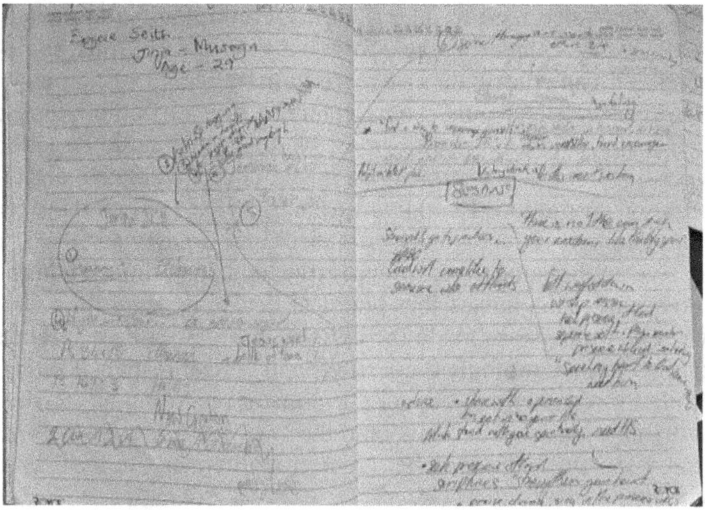

Figure 5.1 Four different examples of Susan Wardell (the author)'s fieldnotes.

Uganda TWO: fieldnotes
Tuesday, 13 August 2013   11:25 a.m.

Rain on razorwire - 29th Dec, 2013

On the stone floor of our living room, a large company of miniscule ants are moving the body of some Olarger insect across the floor with apparent ease, others nearby have discovered the pineapple juice that slipped from our sticky fingers during an earlier indulgence. Outside it is suddenly and deliciously cool as a loud downpour begins. The first rain this month... apparently we have brought the home weather with us. The baby, who has just woken up crying from her place on the huge bed, as the rain began, is now calmly nestled into my shoulder, arms bare in her blue singlet. Nearby Andrew is also silent, appreciating in his own way - with a microphone and iPad, ears intent, making me also listen out for the sounds I now realise come in many layers, as I gaze out the doorway at the torrents of rain on our little compound and the lush red and green Jomayi valley beyond. We are back.

1st Jan

Emotion is being spoken about derogatively here. In casual conversation with myself and Stephen, and in listening to Stephen converse with others, I have noticed it. Emotion is something to be suspicious of, something that may lead you astray, something that cannot be trusted. It is recognised as being a motivating factor, along the lines of a 'needs and desire' philosophy, but to which the moral response is to ignore or overcome. Whether or not there is also a moralisation of self-awareness, as in Christchurch, remains to be seen, but nothing in my work to date would suggest so. In fact that would draw on Freudian psychoanalytics and hydraulic metaphors for emotional management which are strongly at work in NZ, but don't seem to be present here. Instead focus outside of the self, and repression of problematic memories and emotions seems to be the norm.

On preaching, I have struggled more than usual with 'translating' my stories and messages into a Ugandan context. I have a strong sense of the church here being so drenched, so entrenched, in a history and a theology and a reality that I cannot hope to access littleown to overcome. Or perhaps it is more of a fear. My sermon on the Beatitudes, and the 'Upside Down Kingdom', seemed to fall on deaf ears. Coming right after a woman who spoke on the importance of money (earning it, not just being super-spiritual), and in the midst of what seems so much triumphalism, bleeding elsewhere into prosperity teachings, it seemed inaccessible. Presented inside a ramshackle building of wood and tin, where there were no walls to keep out the dust storm that blew through halfway, and cut out our sound equipment, it seemed inappropriate. Bespeckled with terms taken from Switchfoot songs, folk stories, and discussions with my highly educated formed pastor, it seems untranslatable. I neglected (accidentally) to include 'children' on my list of the 'least of these' with whom Jesus associated, and regretted it when later an usher shooed back the children who had raised their hands during the alter call, from the front where adults were receiving salvation prayers. And I cannot for the life of me think of a context I could put the Starfish on the beach story, one of my favourites, into that doesn't involve an ocean that most of my listeners have never, and will never see.

The young woman up the front has her hands up, palms open to heaven. From the raised stage in front of her, the wiry Pastor Samuel is leaning forward to place his hands on her forehead, to grip her cheeks. I see she has a stream of tears down her face, although her expression is peaceful and intent.

Figure 5.1 (Continued)

Shiraz bottle, near the monument to pioneers' first landing

Little monument phone both with blankets and cardboard boxes
Just a couple of hundred metres from margaret mahy playground
A sweeper with Moori tattoo. A waka is docked amidst clouds of smoke from blondes in short denim
shorts and cheetah print tops.
A woman comes up to discuss hiring it for a birthday for 7 and 8 year olds

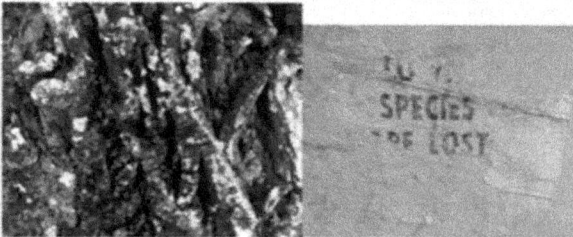

How quietly the water slips under the bridge here
Darkening in a second, without notice or sound
Oxford terrace, a monument to hope
Lions with light in their mouths

Figure 5.1 (Continued)

Figure 5.1 (Continued)

produce. Roger Sanjek shared that during one eighteen-month fieldwork trip Ghana, he generated 397 single-spaced pages of fieldnotes. However, on a different trip, to Brazil, he took no fieldnotes and focused instead on interviews and drawings (Sanjek 1990). Many people recommend writing some fieldnotes every day, which is an excellent goal, but scaling it to what is realistic in your own context is also wise. What is important is that writing fieldnotes is undertaken with care and deliberacy, not seen as supplemental to other more structured kinds of data but as vitally useful to the ethnographic endeavour. This is because fieldnotes are important not just for the product – of a valuable and vivid record of the time and space for your field – but for the process, which can attune your attention, facilitate reflexivity, and help you process the sometimes complex or intense experiences. It is okay to think outside the box about the ways of making notes that will best work for you to do this.

STORY BOX: The texture of the moment

Each day, when Amelia Frank Vitale got back to her apartment after her fieldwork activities, she experienced something worrying: "I could not write". She felt dismayed, even embarrassed, at struggling to undertake this key fieldwork activity. She wondered if it was because of the sensory overwhelm, or the heaviness of her fieldwork in Honduras, which focused on young people who had been deported from the US to neighbourhoods with high rates of violence. After months of struggle, she decided to try something different. Following a fieldwork meeting, she would sit down in her car, turn on her voice recorder, and start talking. Only after a year of being back home in the United States did she listen back to these voice notes. She real-ised they were richly revealing, including "the texture of the moment, the music, my voice". The ambient sounds of the neighbourhood, captured in the background, sparked vivid memories and emotions that "would have been flattened in my notebooks". Through these notes she remembered "what it felt like" to be in that place and with her **participants**, and eventually, it was this that helped her remem-ber how to write.

WHEN AND HOW SHOULD I MAKE FIELDNOTES?

Decisions about how you make fieldnotes should consider three main things. Firstly, what is practical, in terms of recording, stor-ing, accessing, or sharing data later. Secondly, what best helps you capture the vividness and detail of the field and of your own expe-riences in it. Thirdly, the social dynamics of recording, as I return to later.

Fieldnotes are usually defined by their relationship to the space, place, and time of 'the field' (see Chapter 2). Of course, the nature of this field and of what you are doing there can vary

Figure 5.2 'Fieldwork Reflection(s)'. Cabanas, San Pedro Sula, 2019.

Source: Photo credit: Amelia Frank Vitale.

greatly between different projects. While taking notes live is something of an ideal, this is not always possible to do *while* you are participating, either because you can't physically manage both or because it might be awkward or disruptive in that social setting. 'Scratch notes' or 'jottings' are terms to describe the really rough and partial notes that you might take in the flux and flow of things in order to record information or details while they are still fresh in your mind. In the past, these types of notes were assumed to be handwritten, for example in a paper notebook or notepad, but today using smartphones, tablets, or similar digital devices can be equally valid options. The term 'headnotes' refers specifically to notes that go beyond just observational records and include some of your evolving thoughts, serving then as a record of your ongoing processes of sense-making about the field (Jackson 2010). Making fieldnotes is often described as a two-step process, where you jot down as much as possible at the time and then type it up as soon as possible after, e.g. at the end of each day. Even with notes that are already in a digital form, a two-step process can be valuable in helping you to recall extra details and expand on your initial notes, facilitating spaces of reflection and reflexivity.

WHAT SHOULD FIELDNOTES LOOK LIKE?

Fieldnotes are the first step in the "textualisation" of ethnographic research (Lönngren 2021). Generally, they involve some writing, though again this can range from messy jottings to lengthier descriptive or **reflexive** prose. Some people use shorthand (Lönngren 2021). Many people do use excerpts from their fieldnotes in publications later, and these are often lively and engaging in part because they can evoke such a strong sense of 'being there' by drawing on notes taken during or close to the time of the happenings. But there is typically a lot of time and work between writing notes in the fieldwork stage and those published texts. In early stages, it can be better not to get caught up in wordsmithing or preoccupied with style (Wolcott 2008, p. 207), since you need to be able to write freely and record as much as possible in a time-efficient way. That said, fieldnotes are allowed to be written with some verve and flair if this feels natural to you. There is no need to 'dry up' the richness, humour, emotion, or complexity of what you are seeing and experiencing by trying to put on an **objective** scientific voice. You are allowed have a voice, a style, and/or to use creative techniques if that helps you capture the real, rich, **subjective** qualities of the field.

As well as writing, your fieldnotes may utilise visual ways of recording information, including bullet points, lists, doodles, diagrams, or sketches. These can be a quick way to document things that don't translate easily to words. This may be especially valuable in trying to convey spatial information or describe social relationships. Drawing and sketching were, in fact, a staple of early eras of ethnographic fieldwork as an accessible and low-cost way to record observations. They also align well with an ethnographic ethos since they facilitate attention to detail in situated settings and in a way that also continually refers to its own context of production, i.e. the author's perception, their body, their decisions in that moment (Kuschnir 2016). Drawing can be used not only as an alternative to writing, but as a process enabling researchers to see or comprehend new things in new ways (Azevedo and Ramos 2016).

STORY BOX: Drawing the Vā

Vā is an important concept in Pasifika cultures, referring to the relational space connecting people, the environment, and ancestors. Dr Albert L. Refiti is a professor of design and architecture with a background in anthropology. He is the holder of the matai ancestral title Leali'ifano from his grandfather's home of Vaovai, Falealili, in Samoa. He has a long-standing interest in space and relationality and, since 2012, has been the convenor of an interdisciplinary research cluster called 'Vā Moana'. Amidst an extensive Pasifika diaspora, they have explored how the concept of Vā has been adapted widely and begun to influence not only architecture but also policy frameworks. On some of his fieldwork trips, while making observations or listening to people speak, Refiti made sketches and diagrams in his fieldnotes to represent the relational concepts he was trying to understand. He called these 'cosmograms' (see Figure 5.3). He later worked on editing these into colourful artworks that he shared publicly, via Instagram.

Ethnographers engaging with digital, online, or other technologically mediated fieldsites will also have to grapple with how the process of producing fieldnotes may differ in their setting. Screenshotting can be a particularly easy way to create a record of online social activity, since this is already occurring in a textualised/visualised form. But screenshots cannot capture all the things an ethnographer can feel, see, and sense as a participant – the liveness, **temporality**, affective milieu – nor does it really produce *thick* description (as I return to shortly). For all of these reasons,

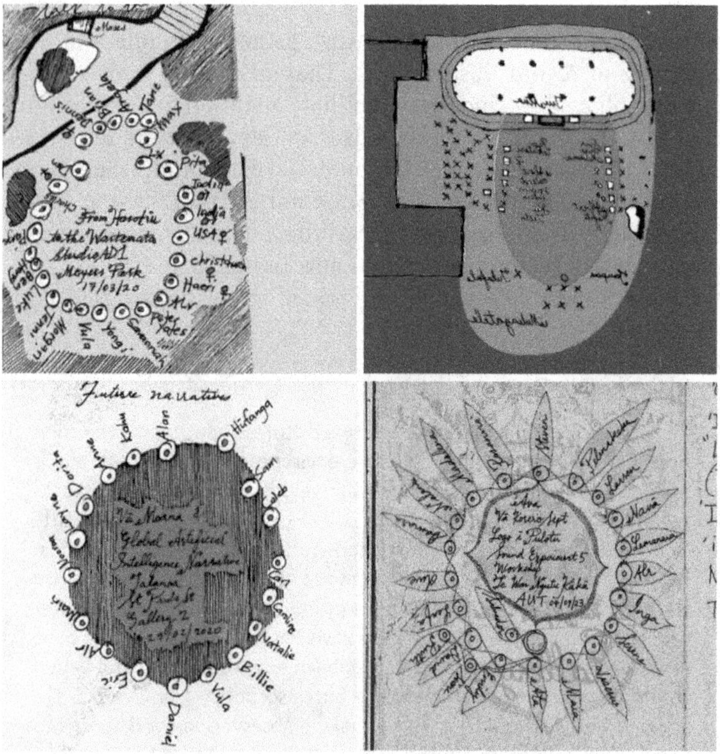

Figure 5.3 Four of Albert Refiti's 'Cosmograms' (2020–2023).

screenshotting should supplement rather than *replace* the process of writing fieldnotes. In fact, the process of writing fieldnotes about observations, experiences, and happenings in that online or virtual space can become a key part of distinguishing digital ethnography from other methods that tend to approach digital sites as *texts* to analyse instead of sites of social life.

WHAT SHOULD I RECORD?

A good starting point when making fieldnotes is to record what you observe. This could include observations about the setting (the physical environment, flora and fauna, buildings, and so on) and about the people (the things they wear, how they move, how they speak). Remember to include not just things you 'observe' in the sense of what you can *see,* but also the things you perceive with your body, and with other parts of your emotional, moral, spiritual, and relational self (see Chapter 4).

This may seem like fairly broad mandate: to record something between 'anything' and 'everything'. Of course, what you record will partly depend on what you are trying to study. But ethnography strives for holism, which means aiming to understand how the different parts of social life interrelate. Because of this, it is important not to limit our focus too quickly by assuming what things are relevant to our research and what things are not. I personally like to invite my students to think of themselves as detectives entering a crime scene, when they enter a field as ethnographers. By this I mean that they have in mind a question they want to answer, but whilst some things are going to grab their attention immedietly, they don't *really* know yet, amidst the ephemera of everyday life, what seemingly mundane things might be clues. As such, they need to observe, absorb, and document as much as possible so that later, as they start to connect the dots, they will have a record of the details that turn out to be relevant. Where the ethnographic process differs from this forensic metaphor is that the goal of creating a detailed record, for ethnographers, is not meant to be achieved in one visit. Fieldwork usually occurs over long periods and/or repeat visits, because a cumulative approach is needed to develop the depth and detail required. In addition, the phenomenon you are recording might not be static but may shift and vary during different times of the day/week/year. This means that temporality, change, and rhythm could be part of what you are interested in, again requiring observations accumulated over multiple visits.

Because of this, another way to think about fieldnotes, is about recording what *happens*. That means people's actions and interactions, things they say and do, events that play out on the social stage around you (and sometimes including you). Since an ethnographer can't be at full attention at every moment (see also Chapter 4), Wolcott (2005) suggests capitalising on shorter bursts of attention to record '**vignettes**'; detailed, evocative descriptions of specific 'episodes' of social life, which might be as short as a few minutes or even a few seconds. Vignettes have often been

Figure 5.4 A drawing by multispecies ethnographer Laura McLauchlan from a session of fieldwork spent sitting in a back-neighbor's yard at night, as part of her study of hedgehog conservation.

characteristic of ethnographic writing. They remind us too that sometimes it is perfectly appropriate to focus in on whatever it is that catches your attention, whilst also being appropriately reflexive about why this might be.

Thick description

It is helpful to think about the overall goal here in terms of the *type* of description ethnographers aim for. This is called '**thick description**'. The term was promoted by Clifford Geertz in the 1970s, and according to Denzin (2001, p. 53), it is characterised by the following:

a   It gives the context of an action,
b   it states the intentions and meanings that organise the action,
c   it traces the evolution and development of the action, and
d   it presents the action as a text that can then be interpreted.

Writing focused on thick description is, because of this, an interpretive act focused on recording not only details but also contexts and meanings. The classic example of this, from Clifford Geertz himself, is a wink. A 'thin' description might be to state, in a factual way, that one side of someone's facial/eye muscles contracted for about X seconds and then released. But a wink can be conspiratorial, cheeky, flirtatious, or many other things, depending on a lot of contextual factors and layered-on social meanings. A thick description will take care to record the who, what, when, where, and why around that action, incorporating whatever else the ethnographer can know or sense about its situated significance in *that* particular field. This is more likely to produce valuable *ethnographic* knowledge.

Recording the personal

Fieldnotes, as a fundamental ethnographic practice, are "distinguished by openness and bravery" (Whittemore 2005, p. 25). Part of this is their work to be reflexive and transparent about the fieldworker's own presence in the field and their role in knowledge production. As such, they should record not just what others said and did but what *you* said and did, with details not just about what you witnessed but how you reacted to or experienced it (Whittemore 2005). There is a long tradition of ethnographers keeping diaries or personal journals in addition to their fieldwork journals or notes. These types of documents provide a 'backstage' space, often capturing some of the emotions and experiences that might be repressed as irrelevant or unprofessional in more 'official' research records. However, there is a disadvantage in separating these things out. As the previous chapter also covered, seeing this as part of the same whole can provide more rich and reflexive insights. Sometimes, this can form a centrepoint of the methodology itself. Indeed, **autoethnographic** studies tend to rely heavily on writing as a technique for producing 'data' about the self as a way to analyse the social world.

STORY BOX: Triggered to write

Emily Darnett is an Aboriginal (Indigenous) Australian woman, who also has blonde hair and blue eyes. When she took up a doctorate in clinical psychology, with a focus on Aboriginal and Torres Strait Islanders' mental health, she found herself reckoning with uncertainties of identity and belonging. About a month after starting her PhD journey, she also began written journal reflections on her own Indigenous heritage and identity. Though she initially attempted to use routinised practices and set specific times to write, she settled on a practice of writing primarily when she was emotionally triggered – when she felt "conflicted, embarrassed, fraudulent, angry or anxious" (Darnett and Rhodes 2023, p. 466). This included documenting her affective responses to specific events in her research process and in the wider world, including the Black Lives Matter movement, COVID-19, and more. Over time, Darnett also talked to other people with dual heritage and reflected on their "similar insights". For her, the written journal formed a route for "inward investigation and recognition" (475) as well as for engagement with the theories of other scholars that "interacted reciprocally" (465) with her own introspection.

WHAT (AND WHO) ARE FIELDNOTES FOR?

Your fieldnotes should be considered less as 'raw' data and more as the first stages of sense-making about your fieldsite. The good news is that fieldnotes are expected to be partial and unfinished data; a form of "provisional knowledge" or "temporarily takes" (Whittemore 2005). Not everything you write is expected make it to your thesis, research report, article, or book, and nor should it. For many ethnographers, fieldnotes can be treated as a reflexive thinking space so you are not constrained by trying to pin down meanings in a tidy way and can stay open to emergent understandings. This said, in some projects, sharing notes transparently or co-producing notes *with* your research subjects can be a useful process, and a way of evening up power dynamics between researcher and subject. Lönn-gren (2021) emphasises a spectrum ranging from 'secret jotting' to 'open jotting', also recognising that what is appropriate and what is ethical may be situational.

STORY BOX: Putting it all on (the) line

From 2008 to 2010, Casey Burkholder – who is a white, queer, Canadi-an-educated woman – taught at a public school in Hong Kong. She also did her master's research focused on this setting. But when she went on to start her doctoral work with her former students in 2015, focusing on the experiences of racialised ethnic minority (discursively constructed as 'non-Chinese' in education policy) young people, she also wanted to critique her own previous practices (2016). She homed on in the prac-tice of taking private fieldnotes, which she saw as "potentially exclu-sionary". For the new project, she created a dedicated digital space in which could publish her fieldnotes publicly. What she put on this site was a mixture of written reflections on emergent themes and visual graphic illustrations that reproduced encounters with participants or stories they had shared using pseudonyms the students chose them-selves. Participants were given a link to the blog and were then able to post comments on the site or share their responses in conversations with her later as part of a process of "member-checking", which she felt better fulfilled the goals of reflexive ethnographic practice.

People are often curious about the ethnographer and their work. However, the process of sharing notes with participants or other stakeholders in the field will still have to take account of potentially sensitive information. Sometimes, information is also of a political nature, dealing with complex legal or moral issues or tricky dynam-ics of power. The ethical responsibility to participants (including that of confidentiality or anonymity if this has been agreed) as well as the ethnographer's own consideration of if and how the informa-tion has the potential to negatively impact the participants, will have factor into decisions about how and where fieldnotes are shared, and in what form.

Figure 5.5 An example of Burkholder's graphic fieldnotes, entitled 'Pork Fu Lam (After)'.

TECHNOLOGIES OF RECORDING

Ethnographers are interested in the vivid, subjective qualities of social worlds and in thinking about what technologies can best capture them. As already discussed, this might include writing and drawing, each with its own unique benefits. But many ethnographers also utilise visual, audio, or other **multimodal** methods of data collecting and recording. Sometimes, these practices can be a considered part of fieldnotes (and can be physically combined into the same document), while at other times, they may be considered separate methods and therefore generate distinct types of record. In this section, I focus on some of the different ways of using these forms of recording in the field, keeping the discussion of how they might be used to communicate ethnographic knowledge to audiences till Chapter 8.

Recording is a metacommunicative act (Black 2017, p. 50). In other words, the act of recording says something, or means something, and has its own way of positioning the ethnographer within their social field. As well as ensuring they follow ethical protocols, ethnographers will have to think carefully about the social effects of their choices about when and how they record. *Will people notice or be aware of you recording? If so, how will this act be interpreted, and how will it affect their behaviour and your relationship with them?* These considerations will be specific to the setting. In some settings, typing into a cell phone or laptop or scribbling in a notebook might not be disruptive at all. In other settings, people may find these actions strange or off-putting. On the other hand, some ethnographers report having accidentally caused offense by *not* recording someone's words or actions, since the process of recording can at times be taken as a symbol of the importance of what is being recorded. Just as an ethnographer makes decisions about who and what to pay attention to (see Chapter 4), they also make decisions about what to record.

Different types of recording may have different advantages or risks. This includes how they affect the researcher's ability to move through the space or engage with participants. While technologies may have a neutral significance in the setting and turn out to be a nonissue, others may have a different impact. In some settings, it might be quite sensitive, risky, or confronting for participants to be recorded with AV equipment, for example. There are usually additional ethical considerations as well – for example, extra permissions needed record in public spaces or to turn on a recording device, even with someone who has already broadly consented to participating in the study. Despite this, while some people suggest technology can be a distraction from or barrier to participation in the field, Pink (2007, p. 175) emphasises that it is simply another *mode* of participation that facilitates different types of sociality and co-presence. Recording technologies can also facilitate different possibilities for producing texts *with* participants, as the following sections unpack.

STORY BOX: Hey, get your camera

For nine months in 2008, Steven P. Black studied a Zulu gospel choir that was "part HIV support group, part AIDS activist organisation, and part performance troupe" (Black 2017, p. 48). A major decision in preparing for his fieldwork was which cameras to bring with him to the South African township. Different cameras offered different affordances for how he could participate in social settings, such as a wedding where the choir was performing or an intimate discussion in the garage where the choir practiced. Though he struggled initially to learn the social rules about when and where it was appropriate to record, his participants also advised him on the use and misuse of technology: "Sometimes group members told me 'hey, get your camera', indicating that they thought a particular moment or event should be documented. [. . .] At other times, a subtle glance or facial expression might cue me to put the recorders away" (59). By the end of the visit, recording had become not only a way to gather data, but a way to engage with people and "an extension of my social self" (53).

VISUAL AND PHOTOGRAPHIC RECORDING

Ethnographers have been recording and collecting visual material for as long as ethnography has existed. Their ways of doing so have evolved as technologies of image-production have developed and based on debates about the ethics and epistemology of images (see also Chapter 8).

In contemporary ethnographic research, the purpose of photography can vary. In the first instance, ethnographers may take photographs of participants, events, or settings that are part of their general fieldsite. Photography can be a powerful way to record either contexts or happenings with a high level of detail. But while the camera has, at times, been promoted as offering something both "more perfect" and "more precise" than the human eye (Kharel 2015, p. 152), ethnographers have had to think critically about the use of visual techniques, given how much scholarship has worked to deconstruct the gaze. All recordings are partial and perspectival; they come into being within particular sets of social relations and relations of power and involve decisions about focus, framing, subject matter, and so on. As such, they can't be seen as neutral or objective ways to capture 'reality'. Instead, the reflexive lens of ethnography positions researchers well to be able to think about photography as a way of constructing a visual text that is stamped with their own point of view . . . and to see this as a potential strength. For example, it makes photography a good tool for autoethnography since it can facilitate "Looking both outwards and inwards" (Gariglio 2023, p. 2). As Luigi Gariglio expresses, "My body, my senses, and my emotions were always involved in the photographic project independently of its focus: whether myself or others" (Gariglio 2023, p. 2).

At the same time, new low-cost, accessible, and mobile visual technologies have made it more possible for research participants to intervene upon the researcher's way of making images. They can work by troubling the directionality of the gaze, providing platforms for *self*-representation and supporting new forms of collaboration. To make the most of this, ethnographers should be attentive to the technologies and forms of visual practice participants already engage in as part of their everyday lives (Dattatreyan 2015).

Visual techniques can be blended with other forms of data collection and questioning (see Chapter 6), as is the case with the 'photovoice' technique. This is a widespread participatory method which focuses on inviting participants to create and share visual representations of their own lives through photography. They then discuss the photos they have taken with the ethnographer, and/or in a larger group setting. The data, in this case, is not just the photograph but the explanation that comes with it and the interplay between the two.

STORY BOX: The picture of health

Emma Mitchell, a student at a school of nursing in the USA, spent eight months (from 2009–2010) with communities in the region of Bluefields, Nicaragua. Her goal was to learn about the health experiences of Creole women, which mainly involved taking fieldnotes and running focus groups. With one such group, she followed up initial conversations by distributing disposable cameras to the twelve participants. Rather than giving the women examples of what to photograph she simply invited them to describe their health experiences through pictures. After developing the photos, Mitchell met with the women individually to discuss what each photo meant to them in a recorded interview. As well as documenting Creole traditions of healing, she found that many had chosen to photograph trenches between houses, wells, water, and garbage, expressing over and over their concerns with environmental conditions as central to the topic of health (Mitchell et al. 2015, p. 29). They shared with Mitchell their wish to get the attention of policy-makers. One participant described the camera as a "powerful" way to fulfil their desire to report was happening in their community. At the end of the project, Mitchell facilitated a photography show in a publicly accessible auditorium, featuring images selected by the women themselves. Though they were disappointed by low attendance of stakeholders at this event, Mitchell also distributed findings directly to key stakeholders later.

SOUND RECORDING

Sound can be an excellent vehicle for both recording and representing ethnographic knowledge. Though it typically gets less attention than image, it too has been part of ethnographic fieldwork since early days. Sound recordings can capture facets of social life that do not have a written correlate or that could only be transcribed in a way in which they might lose their substance (Brooks 2010, p. 622). Some early uses of sound recording were connected to the impulses of salvage ethnography (see Chapter 2). As such, they focused on the task of archiving and preserving certain spoken, sung, or oral cultural forms (e.g. languages and dialects, or local genres of performance) seen to be subject to potential loss, due to colonisation, modernisation, or globalisation. By the 1960s, ethnomusicology was opening out into its own distinct field, "drawing on an ethnographic attunement to sound, and interest in musical and sonic worlds" (Wilson 2017, p. 126). But rich possibilities for attention to sound persisted in other branches of ethnography too.

STORY BOX: Putting Blackness on a/the record

Zora Neale Hurston was an African American anthropologist, writer, and filmmaker and a key figure in the "Harlem Renaissance" in the 1920s and 1930s. Trained by Franz Boas, she travelled widely in the Southern states of the USA and in the Caribbean with the goal of creating an "archive of folk sound" (Brooks 2010, p. 623). As part of her work researching African American song traditions in Florida, she gathered railroad workers, musicians, and "church ladies" to "capture the voices of various informants singing, telling stories, and occasionally hamming it up for posterity" (Boyd in Brooks 2010, p. 622). Using a device the size of a coffee table, sound recordings were made. Hurston's practice used "embodied and sounded performance as a tool of ethnographic inscription, as an instrument that might put black voices on the (scholarly) record" (623). Hurston also sang in many of the recordings herself – and performed the songs later, live, to various audiences, becoming a living archive with performances that blurred the line between archive and technique.

The terms 'sounded ethnography', 'sound ethnography', and 'sonic ethnography' may be used to refer to studies that focus on sound recordings as a main form of data gathering or representation, or that put sound at their centre as a way of knowing (2020, p. 22). Indeed, while metaphors for the ethnographic sensibility have sometimes leaned into an emphasis on the visual (referencing an ethnographic 'lens' or 'gaze'), Clifford Geertz referred also to the "ethnographic ear" (in Samuels et al. 2010). Being able to tune in to sounds through recording and replaying them can facilitate deep listening. Though they offer partial insights, as with any recording, they offer context in their own way, too, since a sound recording will "Always have some representation of the space of performance" within it (Samuels et al. 2010, p. 336). One type of sound recording that ethnographers have found useful is the soundscape, a technique that aims to record the total acoustic environment. Contrary to the 'produced nature' of sound in film, the

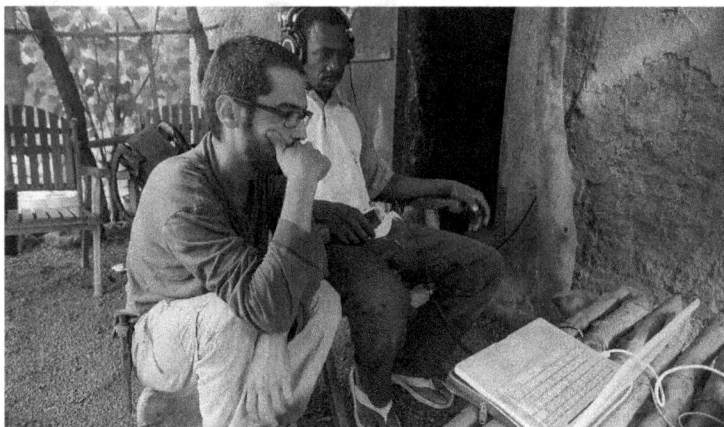

Figure 5.6 (Ethnographer) Lorenzo Ferrarini and (participant/collaborator) Sa-Lasseni Traoré mixing a portion of the soundtrack of the film *Kalanda – The Knowledge of the Bush* together in the field. Karangasso Sambla, Burkina Faso, 2021.

movement towards soundscapes in the 1970s was an expression of ethnographic principles of holism and of **naturalistic** research, capturing the everyday and mundane parts of a social field or setting. The 'sound walk' is another blended, sensuous methodology that draws on the principles of mobile ethnography.

While specialised equipment may allow for better quality, sound recording technology is also now commonly built into smartphones, laptops, and other portable technologies. These provide tools of both recording and editing that have increasingly made it affordable and easy to also create acoustic 'texts' in the field. This can equally facilitate the *co*-production of sound texts, and ethnomusicologist Dave Wilson writes about the "blurry" lines between researchers and informants in this type of practices – for example, when performing music with others, which essentially involves producing a collaborative sonic text in real time (Wilson 2018, p. 129).

AUDIO-VISUAL AND FILM RECORDING

Audiovisual recordings can be valuable for capturing a sense of 'being there'. They can evoke the ethnographer's emplacement in the field through depicting context as well as experience. The process of audiovisual recording as part of ethnographic fieldwork involves questions common to any filmmaker. *Where will the camera be positioned? What will the shot include and exclude? Will it be fixed or moving?* Basic techniques such as framing and angle also communicate, adding meaning to the recorded text. Mobile image technologies – including cell phones, GoPros, and other locative technology – have enabled forms of recording that help explore experiences of place with an emphasis on movement and circulation. This also helps with 'deterritorialising' ethnography.

Margaret Mead believed video cameras could be used "to record thick descriptions of informants and their socio-cultural context through their own voices and activities, based on their understandings of their world" (Kharel 2015, pp. 155–156). Like other visual forms, audiovisual recordings can function to support participatory, collaborative, or participant-driven processes. For example, an ethnographer may ask participants to provide a video tour of a particular space, in which they are able to tell the ethnographer what to pay attention to or notice, thus both showing and telling as part of the same text. Film recordings can also become quite complex

Figure 5.7 Visual anthropologist Briana Young filming in Bali.

co-productions, involving a lot of coordination, collaboration, and planning in a multistep process.

STORY BOX: A forest of voices on film

A Buen Común is a Spanish filmmaking unit with an interdisciplinary background. They argue that film can be an instrument of social transformation. As such members Victoriano Camas Baena, Ana Martínez Pérez, Rafael Muñoz Sotelo, and Manuel Ortiz Mateos aim to use their work to influence policy-makers. One example they share of their strategy of "co-operatively produced" film is a project that focused on the residents of Jimena de la Frontera (Cádiz) and their relationship with the cork oak forest. The actual film shoots happened in two stages in 2000, starting with an initial encounter with the people and place (after a long period of research and preparation), followed by sketching an open-ended script. To start things off they asked participants "What does the cork oak forest mean to you?" and gathered responses from cork extractors, muleteers, entrepreneurs, bank staff, craftsmen, environmentalists, proprietors, local politicians, and hoteliers. The second stage, which coincided with the cork harvest, required two film crews and days of shooting planned in painstaking detail. The process is "time-consuming but rewarding, slow but participatory, and the final work is developed out of everyone's ideas" (127).

Since film is multimodal, it may be able to represent various forms of sensory engagement and entanglement, not just the visual or aural, for example, *showing* people smelling, tasting, or touching. Audiovisual recordings can thus become "routes to multisensorial knowing" (Pink 2007, p. 99).

EXPERIMENTAL RECORDING AND EMBODIED ARCHIVES

Scholars have recognised that the body can be seen as a sort of "organic recording device" in its own right (Madden 2010) – something that can become a repository of memory and meaning. Dance ethnographers in particular have long focused on the body as an archive for cultural knowledge in the form of embodied technique. Pioneering African American dance ethnographer Katherine Dunham established some of the most major contributions to the methods

Figure 5.8 Photograph of Demonstrator Celia Benvenutti, Certified Dunham Technique Teacher, coached by Rachel Tavernier, Master Dunham Technique Teacher, courtesy of the Institute for Dunham Technique Certification, with Harmony Bench and Kate Elswit. For Artificial Intelligence for Creative Movement Analysis and Synthesis in collaboration with "Visceral Histories, Visual Arguments: Dance-Based Approaches to Data".

Source: Recorded at Motion Lab, under the supervision of Senior Creative Technologist and collaborator Vita Berezina-Blackburn, The Advanced Computing Center for the Arts and Design, The Ohio State University. January 26, 2023.

for recording this type of knowledge. In the 1930s, her dance troupe travelled widely, across six continents, though with a particular interest in the Caribbean and African diasporas. Dunham used notational systems to describe and analyse movement as she observed it in different dance traditions and styles. She also recorded film. Even more importantly, however, the knowledge was recorded in the bodies of those who learned the dances and transferred between dance teacher and students over the following generations. More recent efforts to document dance traditions or techniques through motion-capture technologies have fed into crucial debates about possibilities and limitations for centering the body and/or capturing 'visceral' data in the digital humanities (Bench and Elswit 2023).

While ethnographic methods value experiential and embodied knowledge, they also have to continually grapple with the limits of what is describable, recorded, or translatable. The flexible and open-ended nature of the ethnographic methodology lends itself well to trying new things to meet its goal of exploring life as lived. In fact, ethnographers have often been part of experimental work that has helped map the potential of new forms of recording. For example, body-monitoring devices are another avenue for recording information about bodily changes and experiences in the form of quantified or visualised data, and some consumer and UX research studies have used eye-movement monitoring technology as part of ethnographic research to understand how people navigate digital environments. There are many possibilities still to explore, though each requires careful reflexivity around how they might align with an ethnographic lens (see Chapter 1).

USING EXISTING RECORDS

While ethnographers value being able to produce their own records by 'being there', in most settings there will be other records that pre-date the researcher's attention. These may be made by other academics, governments and policy-makers, nonprofit organisations, commercial bodies, private institutions, or media organisations. There may also be records that have been produced by individuals or communities themselves.

Archival ethnography involves identifying, gathering, and interpreting records in a way which applies an ethnographic sensibility and engages with relevant **theory**. It can be used alongside fieldwork or can itself form the main method of a project. From an ethnographic perspective, archives tell you not only about the things they claim to be recording or describing but also about the people, institutions, or processes that produced them. They can be approached not only as texts but as 'sites' in which social life is performed. In a broader sense, as Vidali and Phillips point out, archives are " inherently about relations, responsibilities, voices, bodies, legacies, and publics" (2020). As such, these sorts of secondary sources are no less part of the social field that the ethnographer engages with, and considering how to deal ethically and reflexively with them is just as important.

> **STORY BOX: Violence in the archives**
>
> Salah Punathil dug deep into a "charged" source of information about violence in Kerala in Southern India. Punathil is a sociologist who studies how ideas of violence based on religious difference persist from colonial discourses. One key source of information about the region he studies is reports from official 'Riot Inquiry' commissions. These texts are heavily influenced by colonialist sources and contexts, and are critiqued as a source of information because of this. However he has emphasised an approach that considers not only the content of the reports – and the "sense of everyday life" that they offer – but also ways of using them to analyse "power relations, mediating processes, manipulative moments and bureaucratic performances that make commission report problematic even today" (Punathil 2021, p. 326). Punathil advocates that "Despite all the problematic assumptions in archival sources, they are a potential ethnographic site where the subjects' voices can be heard, allowing the researcher to engage in dialogue with their contents through rereading and reinterpretation" (326).

TYPES OF ARCHIVES

Wolcott argues *"any* document that proves valuable as a source of information can rightfully be considered an archive" (2008, p. 63). Of course, the type of archives available in relation to different topics or fields will vary. "Fieldworkers need to think creatively about available sources of information that are not ordinarily regarded as data, to avoid falling victim to habits that find us invariably gathering the same limited information in the same limited way" (Wolcott 2005, pp. 112–113). The following are some types of records that an ethnographer may wish to gather and analyse in relation to their topics:

Scholarly literature

The first records ethnographers will engage with are often existing scholarly texts about their social field. This may include peer-reviewed academic articles, academic books, and so on. These are considered secondary rather than primary data and, as such, are usually approached through a literature review. There are different types of literature review, and these are a research method in their own right. The goal is generally to learn from what is there and what isn't there, in the scholarly record, as a way to inform what the ethnographer themselves might seek to contribute. There is a benefit to applying a **critical** and ethnographic eye to scholarly texts too, to see them as situated in specific places, times, funding contexts, personal agendas, and so on.

Grey papers

Grey papers are documents with some degree of formality, produced for a specific applied purpose, but that have not been commercially published. This includes things like policy documents, planning documents, reports, newsletters, speeches, plans, court records, patents, meeting minutes, memos, drafts or working papers, and many others. They are often held by organisational bodies, though some will also be on public record too.

Media and popular texts

The media and popular press produce their own descriptions of and stories about people, places, and social issues. They do this in a variety of forms and genres, both fiction and nonfiction (e.g. in movies and television, books and magazines, and news stories) with purposes ranging from information to entertainment to education to profit-making or all of the above. The angle of the reporting or storytelling – what is included or excluded and how things are framed – may be specific to certain social or political trends. The genre or style they are communicated in can also be telling. Since every genre has its own social history, ethnographers can benefit from considering what authority, value, or appeal this style of text is seen to have and why it was chosen for *this* topic and for *this* audience.

Historical archives and personal archives

Libraries, museums, and private collections hold many different sorts of records that go back in time. These can be texts of the sort previously mentioned, or can be more personal texts such as journals, letters, and diaries. Sometimes, participants hold their own extensive archives from their own lives or those of their family members, which they may be willing to give you access to or to show and discuss with you. The use of personal archives in any of these forms can be an interesting approach to autoethnographic work as well.

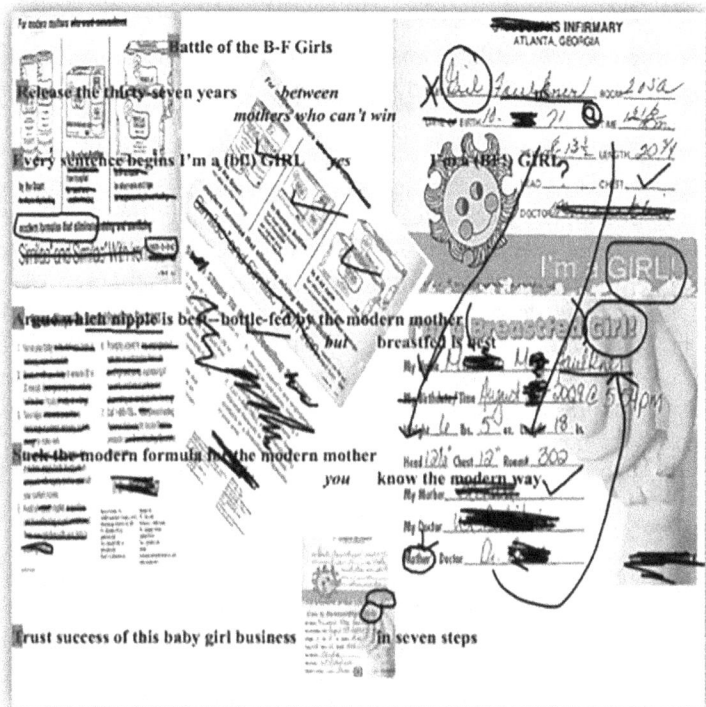

Figure 5.9 Collage from the MotherWork series, entitled 'Battle of the B-F Girls'.

STORY BOX: The material of (queer) motherhood

When Sandra J. Faulker was gifted a damp cardboard box of old family photos and memorabilia by her mother, it provided an unexpected opportunity to apply feminist theory to her own experiences. Combining attention to the material artifacts held in this box with other "systematic recollections", she turned her focus to analysing – and specifically 'queering' – white middle-class mothering using an autoethnographic approach. Faulkner chose to process some of this in the form of several collage or 'scrapbook' poems created from the same personal documents (Faulkner 2017, p. 166).

Digital texts

In contemporary settings, many of the types of text mentioned already will be produced and circulated electronically as well as in print forms. However, there are also some types of texts that *only* exist online: for example, websites, blogs, or social media pages. Digital texts often seem extremely accessible, but it is worth remembering that digital worlds also have spaces ranging from public to private, so not everything is equally visible or accessible to everyone. These records may also include ambiguities and absences – e.g. things being deleted, removed, or censored, links being outdated or broken. Furthermore, digital environments can appear to different users in different ways, sometimes shifting minute to minute and depending on personalised algorithms, making it difficult to 'freeze' them as a static texts.

Literary and artistic artefacts

People record and interpret their own experiences, and perspectives on their own social worlds, in many creative and artistic forms, from poems to sculptures to street art. Some are deemed 'high culture' – often given high status and value, and carefully preserved as such – while others take more ephemeral, popular, or everyday forms. Artistic texts are not exactly 'raw' data, nor are they exactly the same as scholarly peer-reviewed texts, but they provide their own extremely rich form of insight, deeply embedded in positioned subjectivities, while also responding to particular historical, social, political, and sometimes commercial trends.

Material culture

Material culture is also a type of record. A cup, a couch, and a car all form their own sort of archive of social life as well as having their own vibrancy and social agency. In certain cultural settings, particular material forms serve as texts in more specific ways. For example, masks, carvings, clothing, or paintings, in certain contexts, may carry cultural stories or knowledge. Paying attention to material forms as 'texts' can help subvert Eurocentric hierarchies of knowledge, which have historically privileged the written. Seeking to collect and retain physical artifacts is not always necessary or ethical, however. While the public is often familiar with the term 'ethnography' primarily from museums which house 'ethnographic collections', these collections were frequently produced through unequal or exploitative colonial relations. Recent years have seen important debates about ownership and repatriation, which, in turn, have encouraged ethnographers to work on alternative ways to engage with material culture – for example, by photographing it directly, filming people showing or discussing it, or occasionally through new applications of 3D scanning or virtual reality.

TREATING TEXTS ETHNOGRAPHICALLY

Different research fields have their own ways of dealing with the previously mentioned archives and texts. To be transparent and reflexive about their methods, researchers need to be able to explain which texts they have included or excluded from their analysis and why. This is expressed in some fields through the language of **sampling**. While some types of researchers approach this by identifying particular parameters (such as keywords, dates, locations, or publications) with which they might narrow down a set of texts, an ethnographer may take a more inductive approach. To do so, they can draw on their immersion in the wider social field to identify texts that are significant (prominent, recurring, or meaningful) in that fieldsite: *Which posters were on the wall at the doctor's office? What shows were on the TV in the background when you visited a participant at home?* They can also consider asking their participants directly, and in a way which ties back to their research interests (see also Chapter 6): *What apps do they open when they are feeling stressed? What legal documents have they chosen to keep on file? What books do they teach with in their classrooms and why?* In this way, ethnographers can draw on the benefits of their wider participation and immersion in the field to establish a sensitive and contextualised approach to texts, cultivating attention to not only to what information, stories, or meanings the texts contain, but what their role in or impact on people's lives might be.

STORY BOX: A public embrace

In March of 2019, in a town just five hours' drive from my home, a white supremacist killed fifty-one people who were praying peacefully in a mosque. Like many people, I first heard about this tragedy via social media. It was also social media that facilitated a flood of responses from people all around the country – and indeed around the world. When I decided gather some data on this, I combined online participant-observation (generating fieldnotes that combined screenshots with my own observations) and some shorter fieldwork visits to Christchurch (where I also took photos and wrote fieldnotes). I also conducted a series of phenomenological interviews, asking people about their experiences on social media in the hours and days following the attack – what they saw, what they felt, and what they remembered. There was one viral image that recurred in both my notes and my interviews. A Pākehā/white participant told me that she made it her profile picture because it resonated with her experiences of hugging and crying with a Muslim neighbour. A young Muslim interviewee explained how seeing others share this artwork made her feel a little bit safer, even as she hunkered down in her locked apartment. My blend of methods helped me witness and record how this image circulated through social media, and news media, also appearing in material forms at sites of spontaneous memorialisation, working from there to understand the 'why'.

What I consider the golden rule for an ethnographic approach to texts is 'not just *content* but *context*'. By this I mean that ethnographers should not just mine texts for information about other things, but rather treat them as cultural artifacts, whose existence, form, and use is meaningful it itself. This approach is helped by seeing every text as *positioned*. This means recognising it is made by a specific person or people for its own purpose and with its own agendas; produced, circulated, and consumed in association with particular institutions and activities. It is also helped by considering every text as *material*. Many of these types of archives have material forms and, as such, can be treated as both material and textual artifacts. For example, a printed brochure, a badge with a political slogan, and a handwritten protest sign may have the same message but different material mediums or forms. The medium positions the text in relationship with particular bodies and in

Figure 5.10 A variety of instances of Ruby Jones's viral 'hug' image circulating in both digital and material forms.

association with particular embodied social practices. Thinking about what its material qualities indicate about who is engaging with it, where and where, and also what the *experience* of engaging with it is like, can help answer the 'why' of its wider social significance.

CHAPTER SUMMARY: KEY POINTS

- Ethnographers record, produce, and gather data in a variety of different forms towards the goal of creating a holistic picture of their field.

- Fieldnotes are form of text uniquely associated with ethnographic research. They focus on capturing what is observed, what happens, and what is experienced through writing and/or drawing. They are useful for the process (in which they facilitate both reflexivity and interpretation) and the product (of a descriptive and personalised account based on 'thick description').
- Ethnographers may use visual, audiovisual, and sound recording technologies to capture different types of vivid and experiential data about the field. Employing these technologies will shape social relations in the field but can also facilitate opportunities to produce texts collaboratively with participants.
- It is valuable for ethnographers to engage with existing 'secondary' texts, records, or archives related to their field. They may do this in a distinctive way by approaching these not just as information sources but as cultural artifacts.

RECOMMENDED FOR FURTHER READING

Coleman, E.G. (2010). Ethnographic approaches to digital media. *Annual Review of Anthropology*, 39, pp. 487–505.

Elliott, D. and Wolf-Meyer, M.J. (2024). *Naked fieldnotes: a rough guide to ethnographic writing*. Chicago: University of Minnesota Press.

Emerson, R.M. (2011). Writing ethnographic fieldnotes. In: *Chicago guides to writing, editing, and publishing*. 2nd ed. Chicago: The University of Chicago Press.

Kuschnir, K. (2016). Ethnographic drawing: eleven benefits of using a sketchbox for fieldwork. *Visual Ethnography*, 5(1). Available from: https://doi.org/10.12835/ve2016.1-0060.

Pink, S. (2007). *Doing visual ethnography*. London: SAGE Publications. Available from: https://doi.org/10.4135/9780857025029.

Sanjek, R. and Tratner, S.W. (2016). *EFieldnotes: the makings of anthropology in the digital world*. Philadelphia: University of Pennsylvania Press.

REFERENCES

Azevedo, A. and Ramos, M.J. (2016). Drawing close – on visual engagements in fieldwork, drawing workshops and the anthropological imagination. *Visual Ethnography*, 5(1). Available from: https://doi.org/10.12835/ve2016.1-0061.

Bench, H. and Elswit, K. (2023). The body is not (only) a metaphor: rethinking embodiment in DH. In: Gold, M.K. and Klein, L.F., eds. *Debates in the digital humanities 2023*. Chicago: University of Minnesota Press, pp. 93–104. Available from: https://www.jstor.org/stable/10.5749/j.ctv345pd4p.9 [accessed 10 September 2024].

Black, S.P. (2017). Anthropological ethics and the communicative affordances of audio-video recorders in ethnographic fieldwork: transduction as theory. *American Anthropologist*, 119(1), pp. 46–57. Available from: https://doi.org/10.1111/aman.12823.

Brooks, D.A. (2010). "Sister, can you line it out?": Zora Neale Hurston and the sound of Angular Black Womanhood. *Amerikastudien/American Studies*, 55(4), pp. 617–627.

Burkholder, C. (November 25, 2016). "On keeping public visual fieldnotes as reflexive ethnographic practice." *McGill Journal of Education/Revue Des Sciences de l'éducation de McGill*, 51(2), https://mje.mcgill.ca/article/view/9226.

Darnett, E. and Rhodes, P. (2023). Exploration of my aboriginal heritage: an autoethnography. *Human Arenas*, 6(3), pp. 462–477. Available from: https://doi.org/10.1007/s42087-021-00234-x.

Dattatreyan, E.G. (2015). Waiting subjects: social media–inspired self-portraits as gallery exhibition in Delhi, India. *Visual Anthropology Review*, 31(2), pp. 134–146. Available from: https://doi.org/10.1111/var.12077.

Denzin, N.K. (2001). *Interpretive interactionism*. London: SAGE Publications. Available from: https://doi.org/10.4135/9781412984591.

Faulkner, S.L. (2017). MotherWork COLLAGE (a queer scrapbook). *QED: A Journal in GLBTQ Worldmaking*, 4(1), pp. 166–179. Available from: https://doi.org/10.14321/qed.4.1.0166.

Ferrarini, L. and Scaldaferri, N. (2020). *Sonic ethnography: identity, heritage and creative research practice in Basilicata, southern Italy*. Manchester: Manchester University Press. Available from: https://directory.doabooks.org/handle/20.500.12854/63733 [accessed 6 December 2023].

Gariglio, L. (2023). On fieldwork in the hybrid field: a "methodological novel" on ethnography, photography, fiction, and creative writing. *Qualitative Research*, 24. Available from: https://doi.org/10.1177/14687941221149584.

Jackson, J.E. (2010). "Deja Entendu": the liminal qualities of anthropological fieldnotes. In: *SAGE qualitative research methods*. London: SAGE Publications. Available from: https://methods.sagepub.com/book/sage-qualitative-research-methods [accessed 16 September 2024].

Kharel, D. (2015). Visual ethnography, thick description and cultural representation. *Dhaulagiri Journal of Sociology and Anthropology*, 9, pp. 147–160. Available from: https://doi.org/10.3126/dsaj.v9i0.14026.

Kuschnir, K. (2016). Ethnographic drawing: eleven benefits of using a sketchbox for fieldwork. *Visual Ethnography*, 5(1). Available from: https://doi.org/10.12835/ve2016.1-0060.

Lönngren, J. (2021). On the value of using shorthand notation in ethnographic fieldwork. *Ethnography and Education*, 16(1), pp. 60–76. Available from: https://doi.org/10.1080/17457823.2020.1746917.

Madden, R. (2010). *Being ethnographic: a guide to the theory and practice of ethnography*. London: SAGE Publications. Available from: http://ebookcentral.proquest.com/lib/otago/detail.action?docID=743685 [accessed 28 September 2022].

Mitchell, E., Steeves, R. and Perez, K.H. (2015). Exploring Creole women's health using ethnography and Photovoice in Bluefields, Nicaragua. *Global Health Promotion*, 22(4), pp. 29–38, 60, 70. Available from: https://doi.org/10.1177/1757975914547545.

Pink, S. (2007). *Doing visual ethnography*. London: SAGE Publications. Available from: https://doi.org/10.4135/9780857025029.

Pink, S. (2015). *Doing sensory ethnography*. London: SAGE Publications Ltd. https://doi.org/10.4135/9781473917057.

Punathil, S. (2021). Archival ethnography and ethnography of archiving: towards an anthropology of riot inquiry commission reports in postcolonial India. *History & Anthropology*, 32(3), pp. 312–330. Available from: https://doi.org/10.1080/02757206.2020.1854750.

Rabinow, P., Bellah, R.N. and Bourdieu, P. (2007). *Reflections on fieldwork in Morocco: thirtieth anniversary edition, with a new preface by the author*. 2nd ed. Berkeley, CA: University of California Press.

Samuels, D.W., Meintjes, L., Ochoa, A.M. and Porcello, T. (2010). Soundscapes: toward a sounded anthropology. *Annual Review of Anthropology*, 39(1), pp. 329–345. Available from: https://doi.org/10.1146/annurev-anthro-022510-132230.

Sanjek, R. (1990). *Fieldnotes: the makings of anthropology*. Ithaca: Cornell University Press. Available from: https://www.jstor.org/stable/10.7591/j.ctvv4124m [accessed 8 June 2023].

Vidali, D. and Phillips, K. (2020). Ethnographic installation and "The Archive": haunted relations and relocations. *Visual Anthropology Review*, 36(1), pp. 64–89. Available from: https://doi.org/10.1111/var.12197.

Whittemore, R.D. (2005). Fieldnotes, student writing and ethnographic sensibility. *Anthropology News*, 46(3), pp. 25–26. Available from: https://doi.org/10.1525/an.2005.46.3.25.

Wilson, D. (2018). Commoning in sonic ethnography (or, the sound of ethnography to come). *Commoning Ethnography*, 1(1), 125–136. https://doi.org/10.26686/ce.v1i1.4134.

Wolcott, H.F. (2005). *The art of fieldwork*. Lanham: Rowman Altamira.

Wolcott, H.F. (2008). *Ethnography: a way of seeing*. Walnut Creek, CA: AltaMira Press.

6

ETHNOGRAPHIC QUESTIONING, ETHNOGRAPHIC LISTENING

Ethnographers don't just learn by watching. They also learn by listening. Language, communication, and storytelling are key facets of human social life, and so asking questions is a rich opportunity to gain insights into how people make sense of their own lives. But talking with participants always happens as part of a **situated** social encounter. What is said will be shaped by who is involved, where and when the interaction take place, and myriad other social factors. The ethnographer must be aware of all of these things in order to make the most of the opportunity, and to interpret what is said through an ethnographic lens. The chapter starts by discussing the principles of ethnographic questioning and ethnographic listening as they might apply in a variety of different conversational contexts. This is followed by a section on more structured methods of questioning, starting with interviewing and then 'talking in groups', which covers both focus groups and workshops. The final section more briefly discusses elicitation techniques and creative or arts-based approaches that might build on or expand these other practices, in order to make participation more accessible, and generate different forms of data.

ETHNOGRAPHIC QUESTIONING

While the term '**participant-observer**' (see Chapter 4) seems to emphasise what can be *seen*, ethnographers cannot observe everything. They also have to become 'participant-listeners' (Forsey 2010). This means becoming people who ask questions, listen

DOI: 10.4324/9781003404880-6

deeply, and work to make sense of the social nuances of what people do say, don't say, and can't say.

There are many reasons to ask questions. At the most basic level, it can be useful to ask questions about the things you weren't able to observe firsthand or that, by their very nature, *can't* be observed. This might include things that are typically private or unobserved (e.g. toileting, washing, sex), things that can't usually be observed for safety or ethical reasons (e.g. surgeries, dangerous work, sensitive professional activities), or more intangible aspects of social life (e.g. sensations, feelings, thoughts, or beliefs). Of course, not everything can be asked about directly. "Question-asking is culture specific" (Wolcott 2008, p. 62), and asking questions can be rude in some settings or between some people, depending on how they are socially positioned (see also Chapter 3). Sometimes, particular topics are simply out of bounds. But at other times, there are specific *ways* to ask tricky questions. For all of these reasons, "ethnographic question-asking is a special blend of art and science" (Agar 1996, p. 4). Luckily, ethnographic **fieldwork** typically involves a variety of opportunities for talking to people over time and in a variety of settings, and for building this skill while also developing your contextual awareness. This can involve many of the sorts of informal and serendipitous conversations that happen when you spend time in people's everyday settings, but it may also involve planning for a more structured or focused approach, for example in using interviews or focus groups. I start by discussing some of the principles that apply across either setting, drawing on a broader ethnographic lens.

WHAT IS A GOOD ETHNOGRAPHIC QUESTION?

Question asking is a common tool in many different research methods. So what counts as an *ethnographic* question? When talking with our interlocuters, ethnographers benefit from open-ended questions that invite personal stories or descriptions of everyday practices or experiences. Unless it is part of a specific, speculative approach within the project, it can help to avoid abstractions or 'what ifs' (for example, asking people what they *might* do in a hypothetical scenario) and, instead, to ask them about what they normally do or what they did in a specific previous scenario. It is also unlikely to work well to simply ask **participants** a reworded version of your research question(s) (see Chapter 2), which may be much too broad or too theoretical. The goal, instead, should be to get them talking about the things they know best – i.e. their own lives, their own experiences – and then trust the analysis process to help you interpret what these stories reveal in relation to your research questions later (see Chapter 7). The strength of question asking in ethnography is that it doesn't usually occur in isolation. An ethnographer, ideally, over time, builds familiarity with the social world they are studying and with the people in it. Ethnographic question asking can thus also build on what people have shared with you previously or what you already know about the topic in general.

WHAT CAN AND CAN'T PEOPLE ANSWER?

There can be limits to what type of insights we can get from 'reported responses' even in response to the most skilled questions. The first reason is self-awareness. People are often not in the habit of noticing taken-for-granted aspects of their daily lives. There might also be limits to their self-awareness about **subjective** or subconscious things, such as feelings, values, or beliefs. Awareness of specific mental or emotional states or processes is based on socialisation, so in some settings, people may be very used to verbally reflecting on these sorts of things, and in other settings, they may not.

The second reason is that certain things are harder to put into words than others even if you *are* aware of them. The ability to express things in words relies on the words, techniques of language, or narrative forms that you have available, which, in turn, are specific to different linguistic, cultural, or professional contexts. A context in which you observe people frequently struggle to find a way to express themselves around a particular topic also tells you something, as does a situation in which the process of trying to do so is tinged with a particular strong or negative emotion. Indeed, the third reason is that some things may simply be 'unsayable' at either an individual or cultural level, because of trauma or taboo. This can challenge ethnographers to recognise that language cannot and will not provide access to all parts of social experience and that they will also need to listen 'between the lines', as I return to a little further on.

An ethnographer is often inviting people to talk about things that matter to them deeply. This can include painful, private, challenging, or sensitive things. These types of conversations should be grounded in a relationship of trust and an ethics of care, with an eye to wider structures of power and responsibility (see also Chapter 3). It is easy for questions to accidentally trigger harm or discomfort. Again, the ethnographer's awareness of both the social and personal context is helpful for this. Yet it is not possible for an ethnographer to know the whole context, or to prevent *all* emotional triggers. Drawing on Scheper-Hughes's idea of a 'barefoot anthropology', Natalia Luxardo suggests we add the idea of ethnography "on tip toes", "because when entering into these social worlds, like minefields, you never know where the pain is hidden", and respect for privacy is the only way to ensure people are really free to choose to answer questions or not (2022, p. 7). Thinking about culturally specific and culturally safe forms of conversation can also help with creating encounters that feel supportive and secure (Bessarab and Ng'andu 2010).

ETHNOGRAPHIC LISTENING

Listening is a not a passive process but an active and **interpretive** one. It is helpful for ethnographers to think about the social dynamics of presence, attention, and sharing in the encounters they have, or those they are planning. It is also helpful to think about a conversation or interview as a social performance rather than simply an information transfer. Our interlocutors are always representing a certain version of themselves to a certain audience. Ethnographic listening must be ethical and reflexive, factoring in the dynamics of power between the ethnographer, interviewers, or facilitators and those they are speaking with, both to ensure that participants are safe and participating freely, and to consider how these dynamics will shape what is said, by whom, and how.

It may also reflexively consider what the researcher themselves is willing, expecting, or hoping to hear. The researcher always comes with their own preconceived ideas, so **bracketing** (see Chapter 4) can be a valuable practice in ethnographic listening too.

Figure 6.1 Two Belizian university students (centre and right) conducting a street interview with a local man as part of an ethnographic field school in Belize.

Source: Photo credit: Douglas Hume (2019).

LISTENING WITH A GRAIN OF SALT

There can be a big difference between what people say they do and what they actually do. Participants may have their own agendas for the encounter, so an ethnographer can reasonably expect that not all participants will always be 100% honest about everything they share. A famous example relates to Margaret Mead, the anthropologist who conducted ethnographic research about sexuality among young people in Samoa in the 1920s. Despite her fame, an 'expose' in 1983 suggested she had been hoaxed through taking as fact the sensationalised, exaggerated, or untrue accounts of sexual exploits that some of her teenage participants had told her as a joke. This claim has been contested by others since (Shankman 2013). But the fact that it was considered possible for something like this to happen to an ethnographer is interesting in and of itself.

What is important to recognise is that issues with honesty or incompleteness of disclosure may not always be about malicious deception. At a basic level, people tend to want to protect their dignity, reputation, and self-identity. Participants may also be at risk of disclosing things that threaten their safety, are illegal, or are otherwise stigmatised, so suspiciousness or evasiveness may simply mean they are carefully assessing whether you are a safe person to share these things with. Some ethnographers find that over time, participants may deepen revelations about things they have only hinted at earlier or backtrack on information provided earlier once you have earned trust. Ultimately, ethnography draws its strength from being able to place what is spoken and heard into wider contexts, **triangulating** it with other forms of data (see Chapter 7) and applying **reflexivity** around the researcher's own presence as part of the encounter.

LISTENING TO THE NONVERBAL

An ethnographic approach to asking questions is best advanced by paying attention to not only *what* is said but also *when, where,* and *how* it is said. The 'how' means finding ways to document what is communicated in nonverbal ways, for example, with notes taken about gesture, tone, body language. This provides another opportunity for **thick description** (see Chapter 4). Consider medical anthropologist Hannah Gibson's description of an interview she drove three hours to conduct with a woman who "had experienced years of miscarriages and heartache in her pursuit of a family":

> I recorded the majority of the three-hour interview but also focused on making mental notes of important unspoken aspects that could be documented afterwards: The deep timbre of her laughter as she tried and failed to settle her eighteen-month-old baby girl born via surrogacy, who climbed all over her. How her hands managed to multitask without looking, grabbing a drink cup and toy from the floor where the baby had tossed it, as she described the moment she was handed her daughter for the first time.
>
> (Gibson 2019, p. 73)

An approach that combines **fieldnotes** (which may be jotted down right after an interview, if not possible during) with recordings and transcripts can help ensure that what you end up with isn't just a text you are mining for information but a **holistic** record of a social encounter in which the words themselves can be contextualised in other forms of social information from 'being there'.

LISTENING AND TRANSLATING

A lot of what an ethnographer does may be seen as translation between **emic** and **etic** knowledges, or between one cultural frame of reference and another. Some ethnographers also work across more than one linguistic setting, or in multilinguistic communities, translating in a more literal sense. While many ethnographers put strong efforts into learning the language or languages of their participants, at times they may encounter a gap in their ability to engage with and understand people in their **fieldsite**, so there is also long history of ethnographers working with language translators. Translators can have roles in interpreting "not only linguistic but also cultural and inter- and intra-semiotic systems" (Maranhão and Streck 2003, p. xi). This is one reason that the role of the translator has sometimes overlapped with that of the informant or of the research assistant. Care needs to be taken to recognise translation as an interpretive act, and the relationship between ethnographers and translators as another type of situated social relationship. Translators make their own guesses about what the researcher wants or needs to hear. They may also have their own goals and existing relationships that position them in **the field**. Their involvement may shape which participants are willing to engage with you or the study and, if so, what they are willing to share just as much as your own presence.

ETHNOGRAPHIC INTERVIEWING

An interview is a 'conversation with a purpose' (Rapport 2012, p. 53). Since fieldwork can involve many competing demands for your attention, an interview can benefit you by creating a dedicated space to go deeper with a particular person. Unlike the many other forms of conversation you might have, interviews are usually planned and scheduled ahead of time. Ideally, they are also recorded. Because of this, it is usually necessarily to get specific consent from interview participants based on supplying them with information about how their words will be recorded and used and for what purpose and then getting their agreement (preferably written, via a signed consent form) before starting the interview.

An interview, if successful, and once transcribed fully, can be an incredibly rich or dense form of data all on its own. Thus, while some people schedule interviews to supplement data from participant-observation, other people conduct ethnographic studies using interviews as their primary method. This is particularly useful for fields in which participant-observation is not possible or is limited in nature. Importantly, the interview can be seen as an ethnographic encounter in and of itself (Skinner 2020, p. 32), and approached as an extension of **participant-observation** rather than a separate method.

TYPES OF INTERVIEW

Many people have tried to work out what might define an "ethnographic interview", since interviewing is used in so many different types of research (Trundle et al. 2024). Ultimately, the answer is less about the *type* of interviewing and more about some of the broader principles it follows. This section covers a few different possibilities for ethnographic interviewing.

Structured, semi-structured, or unstructured

Broadly speaking, interviews can be categorised as 'structured', 'semi-structured', or 'unstructured'. Semi-structured interviews are considered 'conversational' interviews in that they will involve having prompts, questions, or topics prepared ahead but also leave lots of room for follow-up questions and flexibility according to the natural flow of conversation. These are probably the most common forms of interview used by ethnographers. In some settings, ethnographers may use unstructured interviews, also called open interviews. However, some ethics committees can be nervous about this approach and prefer you to at least indicate *some* examples of the questions you will use.

Structured interviews are those that have a set format with set questions repeated in pretty much the same way across interviewees. Some people count surveys in this category, especially when conducted verbally, though of course surveys can also be done on paper or through digital forms. Surveys are often deliberately succinct so that they can be used with a broad range of respondents in order to gather key information. *How many people live in your household? What activities do you do when you come to this park? Are you for or against a proposed policy change?* While ethnographers aim for **qualitative** rather than **quantitative** data, this process can sometimes provide context for other observations through offering a broader 'snapshot' of the social setting. Questions in structured interviews can be tailored to be more open-ended as well.

In-depth, extended, or repeated interviews

Ethnographers are generally aiming for in-depth interviews, meaning allowing time for the conversation to deepen or unfold is important. An **ethnographic lens** also emphasises, however, that depth is achieved through long-term engagement. Ethnographers may be able to apply this principle by undertaking multiple interviews with the same person. Even just one follow-up interview can be useful, especially if the researcher has since had time to reflect on the content of the first interview and develop thoughtful new, personalised questions from that. Results can differ a fair bit between one-off interviews in the field and repeated interviewing of a person you have worked with over a longer period of time (Skinner 2020), where building of trust, familiarity, and rapport can help with reaching different levels of depth or disclosure. The idea of **'key informant** interviewing' focuses on interviews done with a person who emerges as significant key to understanding in your social field. Some ethnographers also use the method of the life history interview, a technique involving multiple interviews with the same person specifically in order to fill in biographical details and/or map out their experiences over different multiple life stages.

STORY BOX: Interweaving lives in life stories

When Annika Lems moved from Austria to Australia in 2009, she probably didn't expect to gain a "surrogate mother". Yet she experienced a deep resonance with Halima, a Somalian woman and former refugee, who she met there. Both she and Halima had shared the experience of adjusting to a new place, and the fluid sorts of belonging this can generate. By the time they connected, Halima had already been in Australia for eight years. Lems started work on recording Halima's life story soon after in a series of storytelling sessions that spanned eighteen months in total. She wrote later that the act of storytelling initiated a "powerful dynamic" between them. "Through the intimate process of telling and listening to each other's stories, by becoming emotionally involved, our lives began to interweave and mingle, to cross and touch (2016, p. 322). What Halima chose to share over this time came together as "a body of stories that encompassed a lifetime and moved between different places, people and events" (323). Lems also explained that participant-observation formed a "companion" to the recording sessions, enabling her to make connections between experience as reflected upon in stories and "immediate, lived experience" (322). This, in turn, helped her move "away from the idea of life stories as biographical texts, ready to be collected and reproduced by the researcher" and towards a focus on the social *process* of storytelling.

Elicitation techniques, tools, or prompts (as the final part of this chapter returns to) can help interviewees go into more detail or depth. Some researchers utilise interviewing styles that lay out specific approaches to this, e.g. drawing on visual prompts, or material objects, as the main focus of the interview. What might be appropriate depends on the focus of that specific project and on the specific people involved.

Walking interviews, site-based interviews, and mobile methods

Ethnographers often do 'sit-down' interviews in order to create a dedicated or focused space for conversation. Alternatively, however, they may design interviews connected to particular places or activities that are significant to their research topic, providing additional prompts for discussion. Walking interviews, for example focus on visiting a location that has significance to the participant. These are sometimes also called 'go-along' interviews, and they can be particularly valuable for projects interested in place or

Figure 6.2 Hendrik, an eco-creative migrant living in Aotearoa New Zealand, talks to postgraduate ethnographer Yi Li about his artwork while they walk around the garden he developed on the Otago Peninsula, during a visit and interview.

Source: Photo credit: Yi Li (2021).

spatiality, highlighting the idea of the interview as a sensory as well as social experience (Bilsland and Siebert 2024), and further blending questioning and participant-observation.

For the same reason, ethnographers may try to set up other types of field-based interviews attached to relevant activities and in order to effectively elicit information focused on these. For example, Jonathan Skinner described conducting interviews while on the dance floor, "where the close dance personal space" made a difference to what was shared (Skinner 2020).

SOCIAL AND PRACTICAL CONSIDERATIONS FOR INTERVIEWING

For interviewing, practical considerations are entangled with social ones. All sorts of things – from paperwork to recording devices to distracting notifications or background noise – can be unexpected barriers

Figure 6.3 A householder in South London talks the researchers through how he and his partner use their energy monitor during an interview in their home.

Source: Photo credit: Dan Lockton and Flora Bowden (2013).

to (or facilitators of) engagement and thus should be thought about carefully when interviews are planned. The following sections unpack some of this around the practical 'where', 'when', and 'how'.

Where – choosing a location

The location of an interview has to factor in a few different practical things. Firstly, how accessible and convenient is it for you and for your participant/s? Secondly, how practical is it for the task? The public-versus-private nature of different spaces is one key part of this. *What is the subject matter of the interview? How likely are you to be overheard, and will it matter if you are?* Different settings will also create a different tone for your interactions – either more relaxed or more formal. Visiting someone's private residence may be convenient for them. However, there are many different cultural and social norms about visiting someone's home. You are also allowed to consider your own comfort and safety in making decisions about where to meet as well. Meeting in a neutral or mutual location, such as a café or public space, can mediate a few of these factors. But the background noise may not be filtered out well by your recording device, making for lost data, or a difficult time transcribing later. Meeting at a location relevant to the study, as discussed earlier, can also be valuable, but you will still have to factor in accessibility, comfort, distractions, ease of recording, and so on.

Interviews done at a distance – i.e. via a phone call or via a video call – are an established practice. These can be very practical to reduce travel costs and make the process accessible to a wider range of people. They rely, of course, on the participants having access to certain technologies and to mobile networks or internet connections. It can be hard to predict how the mediating factors of something like a phone or a screen will affect the social dynamic. In some circumstances, it may make it harder build rapport, but at other times, researchers have found that it seems to generate a comfortable distance which may actually make it *easier* for people to share.

STORY BOX: Painting pandemic conversations

The pandemic created barriers to meeting participants face to face. This provoked creative improvisation around ethnographic methods (Svašek 2023). Going beyond the limitations of online interviewing, anthropologist Maruška Svašek organised sessions of 'digital hanging out' to investigate how migrant women in Ireland were being affected by the health crisis (Svašek 2023). During these informal encounters, she painted her research participants to visually explore the spatiotemporal conditions of lockdown and of long-distance fieldwork. During the same period she explored using online walking interviews, and photo diaries.

Figure 6.4 'Side by side': a painting by ethnographer Maruška Svašek, which was created during a one-hour online conversation in 2021.

When – scheduling interviews

Different types of interviews need different amounts of time. In my opinion, it is difficult to consider sessions much shorter than forty-five minutes to be 'in-depth', though this may depend on the participant's communication style and on your rapport. On the other hand, an interview may unfold over several hours if the participant and ethnographer are both comfortable and prepared for this. Sometimes, the recording can be started and stopped several times to intersperse food, refreshment, or toilet breaks with conversation. When you set things up with your participants, it is important to flag expectations of how much time you are asking for and to factor in basic needs as well. If you reach the end of the allotted time, check whether they are free to keep going or need to wrap up. You can always consider requesting a follow-up interview if things feel unfinished. On the other hand, if you find conversation with a particular interviewee is a bit stilted and you end up finishing early, take it graciously. There will always be a range of different experiences and outcomes with different interviewees, so don't expect each interview to go the same or put pressure on any one interview to provide all of your insights.

In my experience, interviews are also a very mentally and emotionally intense experience, even (or perhaps especially!) if they go wonderfully well. I try to leave a gap afterward so I can decompress alone, debrief with someone else, or a bit of both, and avoid scheduling multiple interviews in a day if at all possible.

How – recording and transcribing

Most ethnographers will aim to record interviews in some way. To do so you will have to think about what recording device will be practical, secure, reliable, and unobtrusive. Some people are quite used to certain types of technology and won't be too thrown by their obvious presence – a cell phone sitting on a table or desk, for example. Other types of recording – such as video recording, which some ethnographers like to use in order to record and analyse body language and other nonverbal elements – may be more likely to change the participant's behaviour. In all cases, you should signal when you are turning the recording on and off, ideally asking permission to do so even if you already have consent for their participation more generally.

Interviews done in an entirely technologically mediated fashion – including via phone or video but also in written form, for example via emails, Messenger chats, or similar – can be even easier to record, since the recording is somewhat built in. These approaches can also lead to asynchronous interviews in which participants are responding over a stretched-out time period and/or at times convenient to them. This may blend with other forms of participatory methods, including inviting participants to fill out diaries or journals, for example, which can create a space for reflection when they do not feel pressured by time but may still be sharing experiences in response to your own prompts or questions.

Ways of transcribing from recordings can also matter to how these encounters can be interpreted later. Even though it takes a lot of time to produce, having a tidy and complete transcript offers the opportunity for a more structured analysis of a large dataset of interviews: for example were you are looking for patterns in communicative features such as narratives, metaphor, and so on. But denoting spoken exchanges isn't straightforward. The normal patterns of speech are often more complex and messy than we realise. As Emerson et al. explain:

> People talk in spurts and fragments. They accentuate or even complete a phrase with a gesture, facial expression, or posture. They send complex messages through incongruent, seemingly contradictory or ironic verbal and non-verbal expressions, such as in sarcasm or polite put-downs.
>
> (2011, p. 17)

Ethnographers often favour the types of transcripts that denote details about pause, tone, volume, pace, inflection, and so on. A common strategy for example can be to put details about this in square brackets, throughout. This can make transcripts much more time-consuming to produce but also much richer to revisit for analysis later. However, sometimes, it will be suitable to represent someone's comments in a way faithful to the *meaning* rather than the exact words with room later for checking with participants about how you understood, interpreted, or represented their comments.

DESIGNING AN INTERVIEW PROTOCOL

An interview protocol is simply a plan for how the researcher will approach the encounter. These vary in terms of the level of detail or structure. Writing a protocol may be particularly important for research conducted in a team setting, if team members need to establish a standardised approach. However, the process of designing a protocol has value in helping with establishing rapport and flow for individual researchers and those working with more conversational interviewing styles too. Here are some of the things I like to think about in designing protocols:

Introduction and setting the scene

- What key information about myself and my project do I need to include when introducing myself? (i.e. so they feel comfortable and are also fully informed, when consenting).

- Do I need to complete any formal ethics procedures before we begin? (e.g. getting consent forms signed or clarifying their preferences for anonymisation).
- How do I want to communicate the format and style of the interview, set the right tone, and/or establish safe parameters? (e.g. explaining that 'There are no right or wrong answers' or 'I'm really interested in stories about your own life' or 'You are always welcome to skip questions you aren't comfortable with').
- Do I need to obtain any basic demographic info? (If so, what info am I asking for, and will I do this at the start, at the end, or aim to pick it up throughout?)
- Note: always ask if *they* have any questions before you begin.

Questions

- How can I order the questions in a way that makes sense and has a good flow? (With awareness that I will likely re-order them throughout to best respond to the actual flow of conversation)
- How many questions are realistic to this time frame? Which are the most essential ones if I find myself running short on time?
- Are my questions phrased in an open-ended way? Are they clear and jargon free? Are they inviting *descriptive* responses, stories, and personal experiences?
- Are there any questions I need to write out fully to get the wording just right, or it is better to write down general prompts that I can just phrase into full questions more informally in the moment?
- Are there particular questions I can tailor to particular interviewees? What do I need to know about them beforehand, to do this well?
- What follow-up prompts will be suitable to get participants to elaborate or go deeper? (e.g. 'Can you share a bit more about what that experience was like for you?' or 'Would you be able to share an example of what you just mentioned?')

Closing the interview

- How might I structure the interview to end on a positive note? Or at least to safely close the space if it has been a vulnerable or sensitive topic?

- What is the best timing and best way to share my appreciation verbally or to offer tokens of thanks (e.g. vouchers, or personal gifts)?
- What are the key things I need to communicate about when/how I am going to be in contact next or what the future process might be? (For example, for checking transcripts, giving permission for quotes in publications, or potential additional interviews).
- Note: Always ask if there is anything else they would like to share that they didn't get a change to bring up, before concluding – ideally, with the recorder still on.

As suggested here, specific questions can be tailored to the specific interviewee, to what you know about them, and/or to the level of rapport or relationship you already have. In semi-structured interviews a good portion of the questions you ask during the interview should be responsive questions.

STORY BOX: Designing protocols for design ethnography

Ibrahim Mohedas trained as a mechanical engineer, but he soon became interested in user design. In 2013, he travelled to Ethiopia with a team from the United States, working with a company producing a medical device for inserting contraceptive implants. Since the device was intended for low- and middle-income countries, their goal was to learn about the context in which it would be used. On their second visit, they brought the device prototype with them and interviewed fifty stakeholders around the Addis Ababa – including physicians, nurses, biomedical engineers, Ministry of Health officials, FM-HACA officials, and PMFSA officials. All interviews started the same way: with an overview of the interviewee's background and general questions about family planning in Ethiopia. They then demonstrated the device and asked for feedback. Having two researchers in each interview "allowed one of the researchers to consult interview protocols and ensure that no critical questions were missed while the principal interviewer was free to ask follow-up questions". What Mohedas and his colleagues found challenging was the diversity of the stakeholders. "We did not always know a stakeholder's area of expertise [. . .] Thus, we had to prepare ourselves to talk about a range of topics for any one interview", he explained (Mohedas et al. 2015, p. 9). They found a solution in tailoring separate protocols for different types of stakeholders and creating a binder containing copies of each that they could bring with them.

This brings us to the important point that the protocol should be applied *flexibly*, allowing for a natural narrative flow in the interview and for following up on interesting, emergent, or unexpected lines of sharing. As Liza Dalby discovered in her research in Japan, "It was important to leave room for answers and information to creep in sideways [. . .] Often I discovered that what people really cared about was something hovering just outside the categories I had devised" (2011, p. 184). Thus, while preparing questions and protocols *is* valuable, knowing when to set them aside during an interview is also key. In a similar way, making space to reflect *after* each interview and before the next can help you learn and refine your approach and your questions in an **iterative** way.

THE ART OF INTERVIEWING: CO-PRODUCING KNOWLEDGE IN HUMAN ENCOUNTERS

Participants often won't know exactly what to expect when signing up for a 'research interview'. A lot depends on you to model the tone and style of speaking that you want them to take up. Both verbal and nonverbal probes – including things such as 'mmmm' and 'uh-huh' noises that signal you are listening – can really help people to settle into a space of expanded sharing. The communicative patterns in an interview is ultimately a 'zig-zag' (Rapport 2012), where knowledge becomes a "coproduction" between the interviewer and their subject (Skinner 2020). However, the balance between contributors is usually expected to be uneven in terms of both the time spent talking and the amount of disclosure. That doesn't mean there will be *no* talking or disclosure on behalf of the interviewer, however. Interviewer disclosure can sometimes help create rapport by establishing the space as a genuine, shared, human encounter. But ethnographers will have to navigate how much of themselves it is useful, safe, or ethical to put into interviews, and this is extremely context specific.

Interviewees often have their own agendas – a reason that they agreed to this interview and something they want to get out of it. They may simply want to vent. They may need a certain experience to be *heard.* They may want to try and convince you of a particular point of view. It is not uncommon for an interviewee to go off lengthily on a topic that is tangential to your own topics of interest but is clearly important to them. There is a balance to be made here, too, between holding some space for this and gently trying to redirect things. An interview can be quite a big investment of time, energy, and planning. But not every interview will end up being useful to you, and that is okay too.

For all of these reasons, "the interview is an uncertain art or skill" (Skinner 2020, p. 6), and it can be emotionally and intellectually demanding. It requires a type of intense but also split attention. On one hand, you want to be fully present, listening and engaging deeply with your interviewee both as part of the ethical act of witnessing and in order to be able to ask thoughtful follow-up questions. On the other hand, you typically have to keep some attention in reserve for managing time, keeping an eye on your list of questions, making notes about things to loop back to, and so on. It is worth thinking about which tools (i.e. what style of protocols or notes, what type of recording device, what choice of time or setting) might help you to be at your best in this space.

Emotion work and emotional safety

We are often asking people about things that matter a lot to them. An interviewer needs to be prepared for the fact that participants may become emotional or distressed during interviews. If they are handling sensitive topics, researchers should already have engaged with an ethics committee to lay out protocols for managing participants' safety. This usually involves, at the most basic level, paying attention to the subject's emotional state and offering to skip questions, change topics, or end the interview if they become distressed. Offering a break (to finish crying; to get a drink of water; just to breathe) is another option. The most ethical stance I can suggest is to remember that you are a human first and a researcher second. Ultimately, your desire to extract information from the encounter comes second to the participant's well-being. That said, many interviewees *want* to be there, even when it is hard for them, so you can allow for their agency in choosing to do so. In these contexts, listening can be an act of empathy, solidarity, and care.

STORY BOX: Practicing emotional postures

Tam Chipawe Cane, who is trained in psychology and sociology, focused her doctoral research on the experiences of people living with HIV and who were also accessing fertility treatments or adoption services. This involved interviews homing in on very personal and emotional topics. Cane describe how she "rehearsed" sensitive questioning techniques and ways to manage emotions or unexpected responses in a process she describes as "preparatory empathy" (Cane 2018, p. 82). Nonetheless, she found herself deeply affected at times. Her goal, when this happened, was to manage the intensity of her own emotions in order to maintain a "stable emotional posture" for her participants (Cane 2018, p. 82). To help with this, she worked to develop an internal dialogue in which she reflected on her past self and the professional and personal experiences she brought to the research as well as the emotions associated with the interviewing and analysis process. Establishing this practice, she explains, "offered me the opportunity to manage my emotions without causing harm to participants and self in order to formulate a better understanding of participants' life-worlds as they sought to move forward" (Cane 2018, p. 77).

She also found it essential for her own well-being to create space to process the emotions with others outside of the interview. She turned to her academic supervisors for this, as people with whom she could safely share confidential details.

Witnessing people's stories can be profoundly moving and a great privilege. Ethnographers become people who carry other people's stories. We then take on an ethical responsibility for how their words and their stories are analysed (see Chapter 7) and represented (see Chapter 8). However, interviews can also be emotionally weighty for the interviewer. Reflexivity is needed *within* the interview, when an ethnographer should be watching themselves as well as watching the interviewee, since both are involved and affected (Skinner 2020, p. 20). The interviewer often needs to undertake emotion work to balance being present in a human way, knowing that often it is quite okay to react genuinely and empathetically to hearing sad or terrible things and yet not letting their own emotions become the focus. Reflexivity continues *after* the interview in the wider space of the project via planning for the researcher to process their own emotions afterwards, practice self-care, and seek other forms of support and care within professional structures.

TALKING IN GROUPS

Not all conversations or interviews need to be one-on-one. Opportunities for group conversations can come in many forms, both spontaneous and planned. In some settings, including with many Indigenous groups, talking – or "yarning" (Bessarab and Ng'andu 2010) – in groups can be one of the more culturally safe methods of gathering data. At other times, ethnographers may want group conversations to take on more formal or structured qualities so that they can be recorded, transcribed, and analysed. In these contexts, they may become group interviews, focus groups, or workshops. Each provides out a slightly different approach and generates slightly different outputs.

FOCUS GROUPS

Focus groups are a method of gathering data through group discussion and activity. They are particularly common in applied research projects, with a strong history of use in mass communication studies, media studies, and consumer research. They usually home in a particular group of people, sometimes with particular demographic features or who are members of the same community, although they may not be socially known to each other ahead of time.

STORY BOX: Real housewives, real conversations

Chatting about reality television can be a nice break from research. For communications scholar Greta Blackwell, this *was* her research. In 2011, she started a project focused on African American viewers of the TV show *Real Housewives of Atlanta*. She did this mainly through focus groups. But rather than inviting participants to meet her in a room at the university, she set up the conversations to occur in the home of a participant. In her fieldnotes, she documented her nervousness and excitement on travelling to this unknown suburban location for the first time. Her musings on how focus groups might be a "safe space" for people to share diverse perspectives and witness each other blurred into her observations of the "safe" – and, indeed, affluent – neighbourhood she found herself in. Sitting around a glass-top table in the home, Blackwell recorded extensive conversations between her participants about the show. Her fieldnotes added the "experiential knowledge" of being there with them in the spaces connected to their lives. This helped her make sense of "small" details in her data based on the context of their lives as young Black professionals in a predominantly white neighbourhood, which helped her explore "big" topics such as race and privilege (Blackwell 2023).

Focus groups can be even more challenging than interviews in terms of managing time so that each person has space to share, even though more time is typically allotted than for an interview. The group will have its own power dynamics, affecting who has space to share, who feels comfortable and safe to share, whose voice leads, and what experiences or perspectives emerge as dominant or normative. Focus groups need a fairly confident and assertive chair, moderator, or convenor for this reason. The relationships and dynamics in the group will make a huge difference to the outcome, as will the researcher's own presence and experiential data from being in the room and in the conversation. Because of this the space of the focus group can therefore be approached as a fieldsite in itself (Blackwell 2023). The ethnographer can learn a lot by observing (and recording) interaction *between* participants and paying attention to things like language, tone, mood, and affect. *Which topics are unequivocally agreed upon? Where is there disagreement, dissent, or tension?* What is most valuable for an ethnographer, about focus groups, is that you are not just hearing people talk *about* the social world, you are witnessing a piece of it.

WORKSHOPS

Workshops aim to get people talking, interacting, and problem-solving in an even more active and applied way, although the features differentiating a workshop from a focus group can be a bit vague. A workshop may occur over a more extended period of time, for example, over several hours, a day, or even several days. It might involve more than one facilitator. While it can bring together a very specific demographic of people, it may also draw more diversely from a community or population. Workshops often aim to produce a particular outcome or tangible form of knowledge beyond just the transcript of the participants' words. Both workshops and focus groups might use creative or **collaborative** activities as part of this.

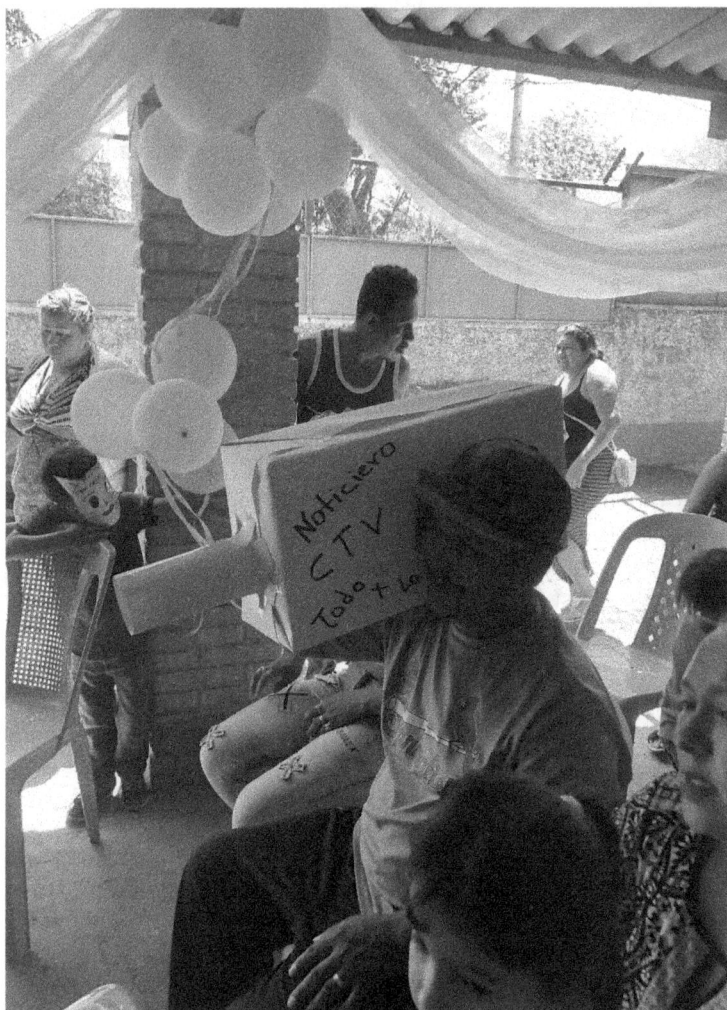

Figure 6.5 Workshop participants perform their sociodramas.

STORY BOX: Workshopping the future

Ciudad Sandino is a densely-populated community on the outskirts of Managua, the capital city of Nicaragua. This community had been settled over several decades by people displaced by various human and natural disasters. Ella Jae Fisher and Alex Nading were part of Proyecto Buen Vivir (The Living Well Project), a long-term ethnographic study in Ciudad Sandino, which aimed to explore how people might both survive and thrive in conditions of urban precarity. Their core approach was a collaborative one, centred on a series of interactive, experimental workshops. The participants were diverse in terms of wealth and education, but the facilitators started by declaring "we are all experts in this room". Groups were posed questions "ranging from the conceptual to the experiential" (Fisher and Nading 2022, p. 7) and the ethnographers' job was to moderate, facilitate, and encourage conversations. Paper and pens were provided. But they were also interested in modes of storytelling based in more local traditions, setting warm-ups in which small groups shared personal histories, national myths, and humorous stories (12–13). At one point, participants acted out 'sociodramas' in small groups, providing the spark for further conversations. The workshops were collaborative and playful and nonetheless had a "political and ethical charge" as personal recollections became vehicles for speculations about the future. Fisher and Nading learned how to ask "the question before the question" and use imaginative engagements to make complex problems more "thinkable" (17).

CREATIVE AND ARTS-BASED APPROACHES

Some people *want* to talk and will talk easily. At other times, people need a bit of prompting. Group settings especially can generate a degree of awkwardness or hesitation around sharing openly. In either one on one or group contexts, researchers can make use of 'elicitation techniques' to help get people talking. Some of this work by asking participants to respond to something more concrete and tangible, like a photo or an object. Indeed ethnographers often find everyday objects are useful as "cultural probes" (Matejko et al. 2021) and can be used as part of something like an 'object interview', as a way of engaging with the materiality of social life (Woodward 2016). Like site-based interviews or walking interviews, these approaches can help the process of gathering data through talking stay connected to embodied experience. The photo-voice technique (see Chapter 5) works in a similar way and can be an excellent strategy to complement other talk-based methods, with a more tangible focus. At other times you might want your participants to respond to some other sort of pre-existing text; discussing, for example a TV commercial, a political billboard, or a website.

Some elicitation technique focus less on a concrete item or text and more on a task in which *they* create something; asking them, for example, to engage in brainstorming, making a list, constructing a personalised map, telling a story, and so on. One direction for this involves what are called 'projective techniques'. These offer specific prompts, stories, or scenarios but ask participants to extrapolate from them, for example, by imagining into the future, either in a specific scenario provided or in their own life-world. Some techniques may work equally well in one-on-one-interviews, and others work best when you have a group who can bounce off each other; it depends on what you are hoping for as the output. These strategies can themselves help break the ice or build rapport, while others are better saved from when you already have a degree of rapport or comfort established.

Many of the techniques previously described open the door for creative or arts-based methods. These allow you to invite participants to express experiences or views in ways that avoid a focus on language entirely, either individually or in groups, where they might co-construct something with the ethnographer and/or with each other. Specific activities might include things like mind-mapping, painting or drawing, scripting and acting out a play, working to develop choreography or improvise music, or any number of other activities. The best results tend to come from thinking about what cultural or pedagogical traditions are already familiar and/or appealing to people in that social context. Fun, joy, and laughter can be valid and valuable parts of all of this, and there is no reason that group encounters in this tone can't generate seriously useful data.

STORY BOX: Drawing a cough with children

Children are sometimes left out of ethnography. They are also left out of important conversations about public health. Julie Spray is a medical anthropologist who works to ensure this doesn't happen. For her doctoral research, she spent time at an urban school in Aotearoa New Zealand with children aged eight to twelve, exploring how they understood risk and managed their bodies with a focus the issue of rheumatic fever. Spray took on an informal role as an assistant teacher in the classroom in order to do this. But she still had to build rapport with the children and work out how to get a sense of their understandings of a topic as multifaceted as health. In what turned out to be a useful solution to both, Julie started not only drawing the children as part of her fieldnotes but drawing *with* the children as she played and talked with them, both as part of group interactions and during one-on-one interviews. She soon found that there were things the children could "explain better through image" than in words. By the end of her project, she had a collection of drawings co-produced through the sharing of drawing materials and the processes of talking whilst drawing, showing the potential for "new ways of co-constructing data" through visual methods (Spray 2021, p. 377).

These sorts of techniques can form part of interviews or focus groups, can represent a modified version of them, or can look like something quite different – that is, a distinctive method or approach (Matejko et al. 2021, p. 7). Used as the main focus of a research project, some of these techniques might be considered valuable for a 'rapid' ethnography (see Chapter 3), since their goal is "to elicit deep, rich data in a relatively brief period of time" (ibid: 8). But they can also work well with groups that return for multiple sessions and are able to develop both rapport and ideas about the creation or output they are working with.

Importantly, arts-based methods can be a route to more inclusive approaches not just because they involve the co-production of texts in some way but also because they may make participation accessible to people or communities who aren't as comfortable or adept with verbal or written communication and may otherwise have been excluded (Matejko et al. 2021). As such, they can help you both ask and answer questions in an entirely different way.

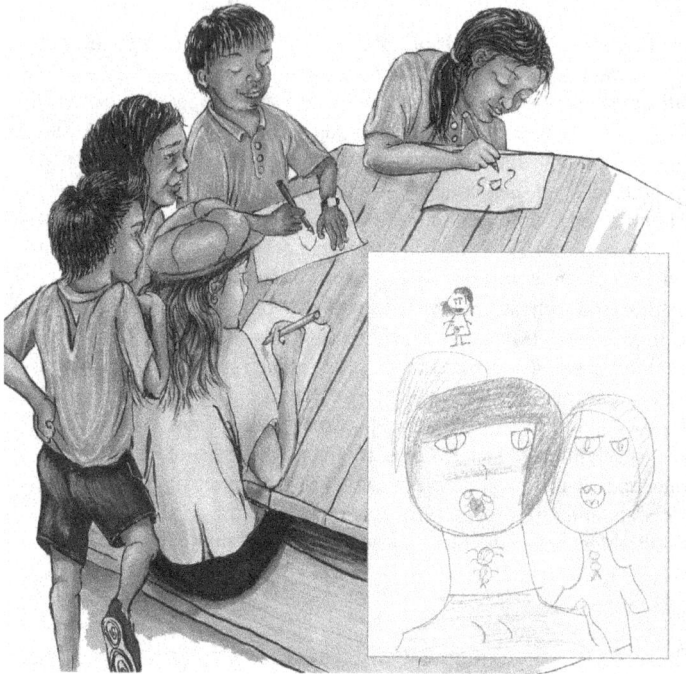

Figure 6.6 One of Julie Spray's illustrations of her ethnographic process of drawing with children at the school lunch break (Auckland 2015); inset image is an example of the type of data this generated: a drawing entitled 'The Sore Throat Bug' by Jordyn, aged 9.

CHAPTER SUMMARY: KEY POINTS

- Ethnographers are not only participant-observers but participant-*listeners*, learning to ask questions, listen deeply, and contextualise what they hear in other forms of social information.
- There are many cultural norms around asking questions and many possible sensitivities too. Ethnographers need to be careful and reflexive around what they can ask and how they ask it as well as considering what can't be answered in words.
- Recorded interviews provide an opportunity for more in-depth conversations and form very rich sources of data when transcribed.

There are a variety of types of interview, each suited to different purposes. An interview protocol can help you think through how to handle the process smoothly.

- Facilitating conversations in group contexts – such as through focus groups or workshops – can enable ethnographers to both listen and observe group dynamics and behaviour, acting as a sort of fieldsite themselves.
- Collaborative, creative, or arts-based techniques can help make participation more accessible and generate different forms of data.

RECOMMENDED FOR FURTHER READING

Bessarab, D. and Ng'andu, B. (2010). Yarning about yarning as a legitimate method in indigenous research. *International Journal of Critical Indigenous Studies*, 3, pp. 37–50. Available from: https://doi.org/10.5204/ijcis.v3i1.57.

Bilsland, K. and Siebert, S. (2024). Walking interviews in organisational research. *European Management Journal*, 42, pp. 161–172. Available from: https://doi.org/10.1016/j.emj.2023.04.008.

Maranhão, T. and Streck, B. (2003). *Translation and ethnography: the anthropological challenge of intercultural understanding*. Tucson: University of Arizona Press.

Skinner, J. (2020). *The interview: an ethnographic approach*. Translation and ethnography. London: Routledge.

Trundle, C., Gardner, J. and Phillips, T. (2024). The ethnographic interview: an interdisciplinary guide for developing an ethnographic disposition in health research. *Qualitative Health Research*. Available from: https://doi.org/10.1177/10497323241241225.

Woodward, S. (2016). Object interviews, material imaginings and 'unsettling' methods: interdisciplinary approaches to understanding materials and material culture. *Qualitative Research*, 16, pp. 359–374. Available from: https://doi.org/10.1177/1468794115589647.

REFERENCES

Agar, M. (1996). *The professional stranger: an informal introduction to ethnography*. Bingley: Emerald Group Publishing Limited.

Bessarab, D. and Ng'andu, B. (2010). Yarning about yarning as a legitimate method in indigenous research. *International Journal of Critical Indigenous Studies*, 3(1), pp. 37–50. Available from: https://doi.org/10.5204/ijcis.v3i1.57.

Bilsland, K. and Siebert, S. (2024). Walking interviews in organizational research. *European Management Journal*, 42(2), pp. 161–172. Available from: https://doi.org/10.1016/j.emj.2023.04.008.

Blackwell, G. (2023). Pursuing ethnographic "closeness": a reflection on race, reality television audiences, and the focus group encounter. *Southern Communication Journal*, 88(5), pp. 416–427. Available from: https://doi.org/10.1080/1041794X.2023.2232808.

Cane, T.C. (2018). Ethics and reflexivity in researching HIV-related infertility. In: Allan, H.T. and Arber, A., eds. *Emotions and reflexivity in health & social care field research*. Cham: Springer, pp. 75–94. Available from: https://doi.org/10.1007/978-3-319-65503-1_5.

Dalby, L. (2011). Japanese ghosts don't have feet. In: *Being there: learning to live cross-culturally*. Cambridge: Harvard University Press. Available from: http://ebookcentral.proquest.com/lib/otago/detail.action?docID=3301024 [accessed 5 December 2023].

Emerson, R.M., Fretz, R.I. and Shaw, L.L. (2011). *Writing ethnographic fieldnotes*. 2nd ed. Chicago: University of Chicago Press (Chicago guides to writing, editing, and publishing). Available from: https://press.uchicago.edu/ucp/books/book/chicago/W/bo12182616.html [accessed 19 February 2024].

Fisher, J.B. and Nading, A.M. (2022). Playing ethnographically living well together: collaborative ethnography as speculative experiment. *Ethnography*. Available from: https://doi.org/10.1177/14661381221083299.

Forsey, M.G. (2010). Ethnography as participant listening. *Ethnography*, 11(4), pp. 558–572.

Gibson, H. (2019). Living a full life: embodiment, disability, and "anthropology at home". *Medicine Anthropology Theory | An Open-Access Journal in the Anthropology of Health Illness and Medicine*, 6, pp. 72–78. Available from: https://doi.org/10.17157/mat.6.2.690.

Lems, A. (2016). Placing displacement: place-making in a world of movement. *Ethnos: Journal of Anthropology*, 81(2), pp. 315–337. Available from: https://doi.org/10.1080/00141844.2014.931328.

Luxardo, N. (2022). Ethics in practice and ethnography: faux pas during fieldwork with structurally vulnerable groups. *Medicine Anthropology Theory*, 9(3), pp. 1–13. Available from: https://doi.org/10.17157/mat.9.3.5747.

Maranhão, T. and Streck, B. (2003). *Translation and ethnography: the anthropological challenge of intercultural understanding*. Tucson: University of Arizona Press.

Matejko, E., Saunders, J.F., Kassan, A., Zak, M., Smith, D. and Mukred, R. (2021). "You can do so much better than what they expect": an arts-based engagement ethnography on school integration with newcomer youth. *Journal of Adolescent Research*. Available from: https://doi.org/10.1177/07435584211056065.

Mohedas, I., Sabet Sarvestani, A., Daly, S.R. and Sienko, K.H. (2015). *Applying design ethnography to product evaluation: a case example of a medical device in a low-resource setting*. Available from: https://www.designsociety.org/publication/37639/APPLYING+DESIGN+ETHNOGRAPHY+TO+PRODUCT+EVALUATION%3A+A+CASE+EXAMPLE+OF+A+

MEDICAL+DEVICE+IN+A+LOW-RESOURCE+SETTING [accessed 5 February 2024].

Rapport, N. (2012). The interview as a form of talking-partnership: dialectical, focussed, ambiguous, special. In: *The interview*. London: Routledge.

Shankman, P. (2013). The 'Fateful hoaxing' of margaret mead: a cautionary tale. *Current Anthropology*, 54(1), pp. 51–70. https://doi.org/10.1086/669033.

Skinner, J. (2020). *The interview: an ethnographic approach*. London: Routledge.

Spray, J. (2021). Drawing perspectives together: what happens when researchers draw with children? *Visual Anthropology Review*, 37(2), pp. 356–379. Available from: https://doi.org/10.1111/var.12244.

Svašek, M. (2023). Ethnography as creative improvisation: exploring methods in (post) pandemic times. *HAU: Journal of Ethnographic Theory*, 13(1), 101–127. https://doi.org/10.1086/725341.

Trundle, C., Gardner, J. and Phillips, T. (2024). The ethnographic interview: an interdisciplinary guide for developing an ethnographic disposition in health research. *Qualitative Health Research*. Available from: https://doi.org/10.1177/10497323241241225.

Wolcott, H.F. (2008). *Ethnography: a way of seeing*. Walnut Creek, CA: AltaMira Press.

Woodward, S. (2016). Object interviews, material imaginings and "unsettling" methods: interdisciplinary approaches to understanding materials and material culture. *Qualitative Research*, 16(4), pp. 359–374. Available from: https://doi.org/10.1177/1468794115589647.

7

MAKING SENSE OF ETHNOGRAPHIC DATA

Ethnographic research is multifaceted, **multimethodological**, and often long term. This means an ethnographer will usually wrap up a project having gathered data in a variety of forms ranging from **fieldnotes** to photographs, recordings, interview transcripts, archival sources, co-produced creative artifacts, and more. Ethnography is also an **interpretive** tradition. This requires researchers to recognise that meaning doesn't simply lie waiting to be 'discovered' in their data, but rather is actively constructed in their own processes of analysis and sense-making, and in dialogue with **participants**. This work doesn't wait till data-gathering is complete to begin. Instead it is engaged with **iteratively** throughout the entire research process. With all this in mind, this chapter offers some insights on the processes of interpreting, analysing, and theorising from ethnographic data. It starts with a few basics about how to judge 'enough' data. Then it moves into thinking about a structured and practical approach to organising, categorising, and **coding** what you have and beginning to look for patterns. Even bigger questions come in around the conceptual tools ethnographers use to produce theories about the social world, with an emphasis on context, comparison, and **critical** analysis. In thinking about the relationship between ethnography and **theory**, the chapter taps into both **epistemological** and representational issues, highlighting the **collaborative** nature of ethnographic sense-making and the continual **reflexivity** needed around this process.

DOI: 10.4324/9781003404880-7

WHEN DO YOU HAVE ENOUGH DATA?

Ethnographic data is typically partial, provisional, and open-ended, something researchers build over time, sometimes through work that continues with the same communities over years. Nonetheless, practical limitations will most likely mean you need to wrap up formal periods of **fieldwork** sooner than you'd like. You will also want or need to publish some findings eventually. One of the anxieties of conducting research can be determining how much data is 'enough' to do so. In other words, when are you 'done'?

In part, this can be about feeling able to answer your research question. This doesn't mean establishing an authoritative or final answer but rather being able to respond to the question with some degree of detail, nuance, and rigor. Of course, your ability to do this will depend on how broad your question or focus is, which is one reason we have to scale our questions to the time and resources we have available (see Chapter 2). But determining when you are done can also be helped by thinking about theoretical saturation, 'rich' data, data **triangulation**, and immersion.

DATA SATURATION

Data saturation refers to a point at which new data coming in does not offer any new revelations on the topic. Ethnographic data is **situated** and idiosyncratic, so this won't mean no new stories are coming in at all but, rather, no majorly new *theoretical* insights or ideas are being identified within them – something more specifically referred to as 'theoretical saturation'. While reaching saturation can be a useful measure of validity (Aldiabat and Le Navenec 2018), it can also be a challenge for small-scale projects to plan around this, as there is no clear-cut way to predict how much data (i.e. how many interviews conducted, how many hours spent on **participant-observation**) you may need to reach this point. Like everything else, it can be rather subjective, and the researcher makes their judgement as best they can within concrete parameters such as funding deadlines and timetables. Of course, as they begin to present or publish their results, they will need to be transparent about the scope of their data gathering, so it will also be open to the scrutiny of others. Feeling insecure about this can lead to putting off analysis or writing by fixating instead on getting 'more' data, but Kleinman and Copp warn against "compulsive data collection" (1993). There is rarely a *definitive* stopping point, but it helps to remember that ethnographic research is more often evaluated by the richness of data than by the amount of data.

RICH DATA

We talk about data as 'rich' when it is descriptive, multilayered, detailed, and situated. This is part of the reason you can't define data saturation in purely numerical or **quantitative** terms. Achieving rich data is partly dependent on the skill of the ethnographer and partly dependent on the responsiveness of the participants. Some forms of data also tend to be richer than others, meaning, for example, that researchers may find fifteen interviews more rich than 500 surveys. Richness also comes from having a variety of different sources of data so that you can triangulate between them.

DATA TRIANGULATION

No single method, theory, or observer can capture all that is relevant or important (Denzin 2017). Instead, using multiple methods and connecting between the insights that each provides can help explore a social **field** or phenomenon more rigorously and holistically. The process of connecting between multiple different forms of data is sometimes called 'triangulation'. Triangulation can help you get to data saturation more quickly as well as ensuring depth and rigor in your analytic process. Thus, to an ethnographer, a study that includes a modest number of interviews, some fieldnotes from participant-observation, and analysis of a small set of relevant archival texts might be seen as offering richer insights than a study comprising a vast amount of just one of these forms of data. The process of connecting and comparing between different parts of your data may happen somewhat intuitively as you proceed with the study, but it can be aided also by a structured approach to analysis, as I come to shortly.

DATA IMMERSION

In Chapter 4, I discussed the idea of 'immersion' in the field as a goal for ethnographic research. Analysing data can be a way to extend and deepen immersion. For example, some ethnographers find that the process of transcribing interviews, though often time-consuming and laborious, is invaluable in giving them time to deeply consider, at a less-pressured distance, the nuances of what was said and how. Coding (as discussed shortly) can also help with immersion, as it means returning attention to textualised forms of data in close detail, often several times over, as you listen or read line by line.

Care is sometimes needed when you are immersing yourself in data that deals with sensitive, traumatic, or troubling content for long hours. 'Data' in ethnographic research is not something cold, hard, and abstract, after all, no matter how the word sounds. Rather, it may focus on descriptions of poignant experiences that participants or the researcher themselves has lived or witnessed. It may include rich sensory data. It may include things that were confusing, uncomfortable, or unresolved. Reconnecting with the emotional aspects of this can be valuable rather than something to simply push through, as this, too, can enable reflexive insights. But it means the process of data analysis can be just as emotionally demanding as other parts of the research process.

ITERATIVE PROCESSES

For ethnographers, gathering data and analysing data are not distinct processes. While there can be a tendency to assume that these will be two separate steps – with data gathered first in the field and then analysed later back at home – for ethnographers, analysis *starts* in the field, and proceeds in an ongoing way. Some insights might emerge most poignantly amidst the vividness and urgency of lived experience, while other things may become clear only later, with an opportunity to take a more distanced, structured, or methodical approach to reviewing what data you have gathered. This is why ethnographers ideally move back and forth between gathering and analysing data throughout the project. Where there do turn out to be gaps in the data or where analysis starts to show areas the researcher might like to know more about, an **iterative** process can allow for returning to engagements with participants and with the field armed with even more focused questions.

ANALYSIS AND INTERPRETATION

Just as there are different methods for collecting data, there are different approaches to or techniques for analysing data. An **analytical framework** establishes what to do with your data. It prescribes a particular way of deconstructing what you have and looking more closely at certain aspects of it. Some analytic frameworks have been developed within a specific discipline, while others are used more widely. Examples of analytical frameworks include thematic analysis, content analysis, discourse analysis, narrative analysis, semiotic analysis, and framing analysis, but there are many more. Each has its own established protocols and strategies, and each helps highlight a different facet of the data or highlight it in a different way.

STORY BOX: Ten years to unpack

Julia Davies is a Pākehā – a New Zealander of European/settler decent – who undertook an autoethnographic project exploring her relationship to the Indigenous Māori healing practice of rongoā. Her main form of data was a personal archive of journals spanning ten years. The journals provided detailed accounts of emotions and internal dialogue as well as experienced and observed events. Davies recognised that the journal entries involved a great deal of sense-making she had already been undertaking about her own life, and thus, her research did not involve starting analysis from scratch but rather applying a more structured approach. Towards this goal, she chose to apply a framework of narrative analysis, which, as she explains, "provides an opportunity for **autoethnographic** writers to facilitate a connection between their story and the understanding of **culture**" (Davies 2023, p. 26). After identifying sections of the journals that were particularly relevant to her research question, Davies focused on establishing a chronological account of her journey. As a next step, she looked for "patterns surrounding feelings, emotions or situations before any healings or spiritual experiences" (25). To do this, she followed a process for narrative analysis which broke things down into distinct steps so that she had a structured process to follow.

For certain projects, an ethnographer may select and apply one main analytic framework. This can help to create a clear and consistent way of handling the data. However, the multimethodological,

holistic nature of ethnographic research often lends itself to more of an eclectic approach to allow for moving through multiple layers of analysis. As such, many ethnographers find it useful to deploy a variety of different analytic techniques and tools for different parts of their data, or in combination. It is also very much a case of building the plane while you are flying it in that ethnographer's don't typically devise analytic protocols ahead of time – at least not in full detail – but let them emerge, inductively, "after the fact" (Strathern 2004, pp. 5–6).

Furthermore, while analytical frameworks lay out protocols and procedures, analysing ethnographic data can also be an "intuitive, messy and serendipitous" task (Pink 2015, p. 141), drawing on the researcher themselves as an interpretive tool. There is value, as a first step in being able to sit with data in a more overarching and **humanistic** way. *What stood out? What felt unexpected or surprising? What was a recurrent and persistent trend? What paradoxes or contradictions emerged?* This can involve listening for the gaps – for silence and absence. It can involve focusing on the intangible and the tacit. It can mean asking yourself what matters to people, or what is most at stake in their social worlds. Returning to these bigger sorts of questions at regular junctures throughout the process can help counter the tendency of some forms of analysis to produce a **reductive** way of thinking about the field, and thus avoid 'analysing away' the lived quality or the wholeness of human experience.

ORGANISING DATA: CATEGORISING AND CODING

Analysing ethnographic data in a structured way can be complicated by the multiple forms that it often comes in and the volume of what can accumulate over a long period of fieldwork. The first step is usually to sort and organise what you have. Organising data isn't just about organising files but rather starting to organise *ideas.*

There are many different ways to organise data, and it isn't so much about picking one as about exploring the possibilities of each. For example, organising data around some sort of chronology can often be relevant. This might be the order you collected it in, or the timeline or history of a particular person, institution, or group. Another option may be gathering up all the data pertaining to a particular 'cases' within your research – where 'cases' may be individual participants, families, organisations, places, or sites. Descriptive categories are also key, and can be the first step in starting to interpret findings, enabling you to show connections *between* different data parts of the data and build towards identifying key concepts or themes. It all comes down to what you are trying to understand through the data and trying different ways of arranging or labelling things might reveal different connections.

STORY BOX: Analysing organisational change

The defense industry is a morally and politically fraught arena. It is also an everyday workplace for many people. Political scientist Dana Landau and organisational ethnographer Israel Drori conducted a three-year ethnographic study of an R&D laboratory in Israel in the late 1990s (2008). Their focus was on how those working in 'NuLab' made sense of change within the organisation at a time when government funding cuts were leading to major restructuring. To examine this, they used participant-observation along with a total of 131 semi-structured interviews, including a cross-section of different staff, both new and veteran, and including scientists, managers, and administrators. For the first step in their analysis, they triangulated between the different forms of data by compiling them all into chronological accounts of the particular organisational crises NuLab had faced. From there, the two authors worked separately on identifying emergent categories before coming back together to compare their results. Next, the data within each category was examined for patterns. Later, the data was reorganised differently yet again, according to the different stakeholders in the organisations. Through this, they were able to show how managers versus scientists, for example, had different ways of making sense about their jobs, their workplace, and the changes in it. For additional reliability, they presented their main findings to **key informants** at NuLab, "incorporating their comments and feedback to our analysis when we found them useful and insightful" (706).

Coding

Coding is a common tool for organising and analysing data in **qualitative** research projects. There are many different types of qualitative coding, but in general, it refers to a process of attaching a label to something to describe its content or characteristics. Skjott Linneberg and Korsgaard describe it as a way to do an 'inventory' of your data (2019). Coding can be done with interview transcripts, media articles, web pages, or anything else text based, including your own fieldnotes. It can also be done with visual, audio, or audiovisual material. Coding different forms of data with the same set (or overlapping sets) of codes can help with triangulation. Software – such as NVivo, ATLAS, MAXQDI, or similar – can support this, and these tools are especially useful when you have large amounts of data, though the same process can be done manually for smaller datasets.

Usually, coding involves working through methodologically – e.g. reading line by line, start to end. A code can be applied at any level or size unit, meaning you can code a single word, a whole paragraph, or a three-page-long narrative ramble. Some parts may need to have more than one code attached.

There are many different approaches to coding: for example, thematic coding or open coding. Broadly speaking, **inductive** coding describes a process in which you build a list of codes as you go based on what you notice in the data. On the flip side, with **deductive** coding, you start with a pre-defined set of codes. Abductive coding describes a process which blends the two, and in my experience, most ethnographic coding tends to fall into this middle space. Most people use some form of codebook or code list to keep track of their codes and denote how each is being applied. This is also a useful way of making your process transparent, with researcher's sometimes including their codebook in articles, reports, or other research publications. The codebook can also be particularly important in a collaborative or team project. In fact, when a project is going to rely heavily on the outcomes of coding to determine its findings, it is desirable to have more than one person involved in the process so that they can undertake inter-coder checking. This means they will code some or all of the data independently from each other and then compare the results. 'Inter-rater reliability' is often used as a measure of validity.

Importantly, the coding process is also iterative. Often, you may have to split codes that turned out to be too broad into smaller

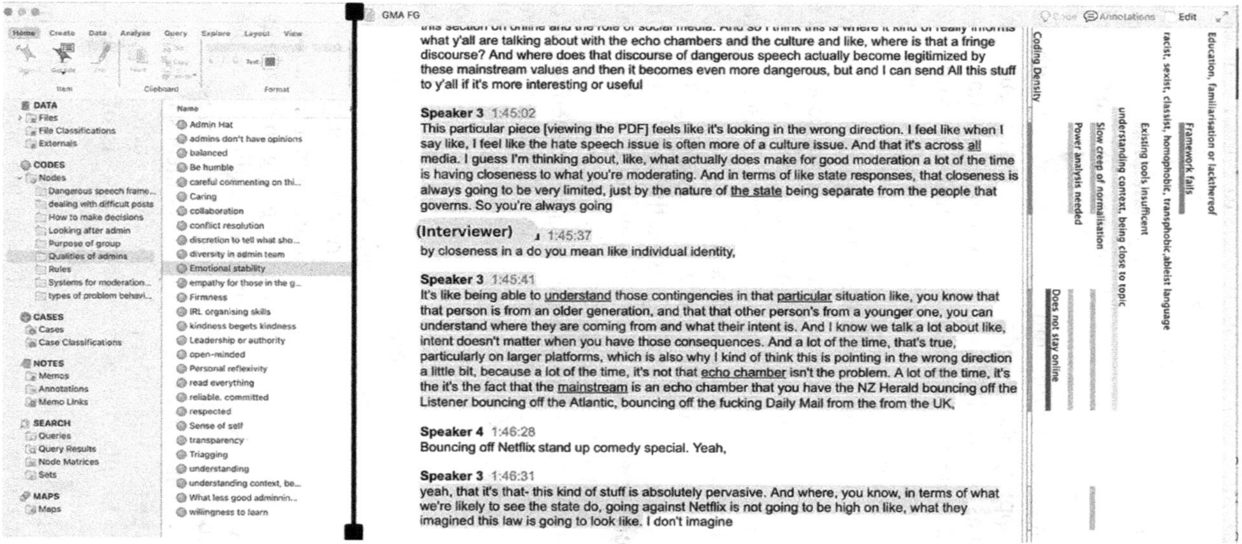

Figure 7.1 Left: Part of the code list for an applied, team-based project led by Susan Wardell (the author) and focused on social media admins and their responses to dangerous speech. Right: segment of a focus group transcript coded on NVivo.

categories, join codes that turned out to be too small or too similar, or create more than one layer of (primary and secondary, or 'parent' and 'child') codes. These steps can help you think through the relationship between the different ideas you are working with. In this way, coding and interpreting are not two distinct phases but "interrelated processes that co-evolve" (Skjott Linneberg and Korsgaard 2019, p. 266). Using memos – that is, "written notes that elaborate what codes mean, how they link, what they start to say about the data, and how they might be developed into a more coherent argument (or grounded theories)" (O'Reilly 2012, p. 204) – can help turn coding into a reflexive conversation with yourself or your fellow coders. Ultimately, for ethnographers, coding "provides structure and depth to the analytical process" (Skjott Linneberg and Korsgaard 2019), but ideally remains an interpretive rather than rote process.

IDENTIFYING PATTERNS

Processes like coding generate descriptive categories. But these are not findings – at least, not yet. If the first steps of coding are like organising your data into baskets, each representing a different topic, the next steps are about taking a look into each basket in order to see (and try to understand and describe) what is in there, or in other words, what you have found out *about* that topic. Even more crucially, findings come from looking at the relationship *between* different codes or data points, which can be connected to start to form themes.

Looking for patterns is a key part of all of this. This may mean looking for overall patterns in the dataset, looking for patterns over time, or looking for patterns specific to particular places, activities, or people. These patterns can help you analyse data that otherwise appears quite diverse and idiosyncratic at the level that most ethnographers are thinking about, that is, i.e. with a focus *shared* identities, *group* norms, and so on. Paying attention to things that seem to *break* a pattern – for example, words, actions, or stories that seem like exceptions to some kind of (typically unspoken) rule – can be surprisingly useful too. Again, this can be helped by the researcher following the sort of intuition or hunches which are informed by their wider ethnographic immersion, rather than being rigid about sticking with the "certainty" of codes (Bönisch-Brednich 2017). This links back to the idea of continuing to ask the more overarching and human questions, as you proceed.

Interpreting ethnographic data is complex intellectual labour. Some of the process is invisible; it takes time and doesn't always feel 'productive', since sitting and reflecting deeply doesn't generate specific outputs. Recognising and valuing the behind-the-scenes, intuitive, and 'slow' processes that can generate deep insights, is helpful for ethnographers. On top of this there are tools and techniques that can help you to do this thinking in a more active way.

WRITING TOWARDS INSIGHT

Analysing data requires time and multiple cycles of thinking, writing, and processing. For ethnographers, writing is not just the *product* but part of the *process*. It is a reflexive practice that helps you establish a conversation between yourself, your data, and your fieldsite. Overall, if you start writing early, you can expect it to generate new insights rather than just lay out those you have already established. Because of this, it can help to start to writing well before you feel ready. Annotations, memos, or regular journal entries can be ways of doing this throughout the research journey. Letting yourself move into more free-flowing writing without constraint or too much focus on writing conventions, tidiness, or quality, is valuable for letting ideas simmer to the surface. Alternatively (or additionally!), setting yourself little writing challenges can also force new clarity. Especially helpful are exercises that require conciseness – i.e. summarising a particular emergent theme in one paragraph, or describing the connection between two codes in just one sentence. If you want to really challenge yourself, try answering your research question/s in just two sentences, even if you are still in early stages of analysis. Coming back to this exercise regularly can help you track your thinking as it develops.

SEEING AS UNDERSTANDING

STORY BOX: Visualising healthcare journeys

Over half of the world's population cannot obtain essential health services. In 2021, the ARC project team, drawing on collaborations with local partners in Bangladesh and South Africa, combined ethnography and design research to speculate about what a more desirable healthcare future might look like, based on the experience of current healthcare users and providers. Their first goal, when looking at the data they obtained, was to identify and categorise different types of frictions people had experienced while trying to access healthcare. They used a digital whiteboard to arrange findings from both countries according to key themes that had emerged and to identify which were common in both settings, and which were more prominent in one than another. They organised particular sets of data (such as ethnographic video interviews) into themed playlists. Based on immersion in this data, they theorised six distinct strands of health. Visually mapping individuals' health-seeking journeys provided another way to compare key features by breaking the journeys into parts and identifying key inflection points, such as diagnosis (Analysis and Synthesis no date). This also enabled them to apply the six-strand framework in order to analyse data at higher level.

Visual approaches to analysis can be extremely useful. This might involve arranging your data into some kind of mindmap or diagram that uses colour, shape, or the spatial relationships of the page to help you identify patterns or relationships between different ideas. Many dedicated digital tools can help with this and may also help you produce something that you can actually share or publish later. However, the tactile, embodied, and 'low-tech' process of scribbling on a whiteboard or a large piece of paper, or rearranging sticky notes on a wall, can feel different from working on a screen, and in my experience can be even more ideal for playing with ideas at an early stage.

cannot be left alone at home. Her husband and sons have to manage their work routines keeping this in mind. She requires constant support with all household activities.

Social Health

serious problems right now."

Physical Health

Living with a chronic disease burden

For Tasneem who suffers from diabetes and a number of related ailments, the strands of her health have been stressed over time creating a challenging living environment for her and her immediate family.

With her husband losing his steady job some time ago, the financial health of the family deteriorated severely, impacting Tasneem's ability to properly care for her physical health. The family's social health deteriorated too, when due to financial losses, they had to move from where they had been living to a new community in Korail - a slum in Dhaka. This move crystallised the downward movement down the social ladder that Tasneem and her family experienced, eroding their identity, confidence and sense of self worth.

Through our structured interviews with Tasneem and her husband, we inferred that the family falls back on their spiritual health to cope with the stresses on their other health strands.

She has known about her diabetes and kidney problems but has no been able to seek treatment for th because of their current financia situation. This has further worsen her condition.

Financial Health

Trust in this religious man seemed to emerge from his refusal to accept money in return for the healing. This is in sharp contrast to the transactional nature of other forms of healthcare

Financial Health

Tasneem's family falls back on their spiritual health to cope with the stresses on their other health strands.

Emotional Health

Spiritual Health Seeking

In an interview with us Tasneem shared an instance of health seeking that occured a couple of decades ago when her daughter was a young child. At the time Tasneem and her family used to frequently seek treatment from a religious man.

She went to him when she was having difficulty conceiving, and to cure her daughter when the child could "see the jinn". According to Tasneem, this healer was able to provide solutions to all the problems that her family approached him for and insisted that his healing powers had helped others, who had suffered from more severe health issues.

For Tasneem, who is apprehensive of ayurveda and other kinds of alternative treatments and medicine, there was a certain purity to this healer who she trusted fully. Part of that purity seemed to emerge from his refusal to accept money in return for the healing. This is in sharp contrast to the transactional nature of other forms of healthcare including modern medicine but also other practitioners of alternative medicine/healing.

Figure 7.2 One section of a visual map of the health-seeking journey of a diabetes patient from Dhakar, Bangladesh.

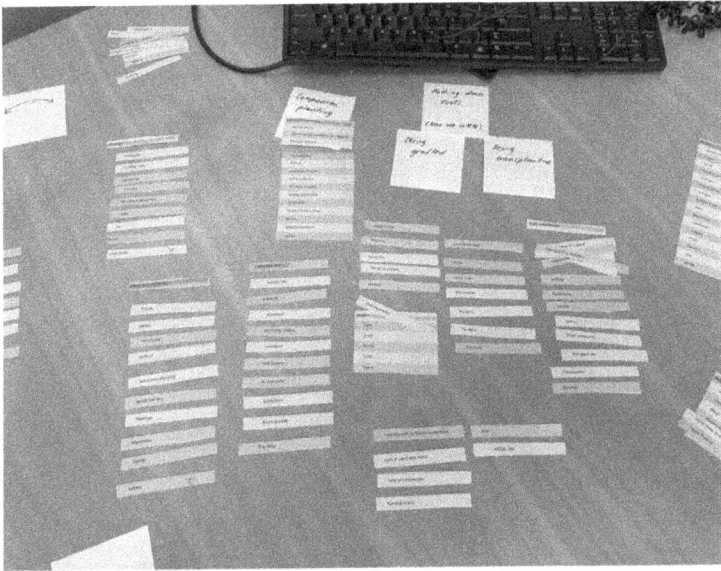

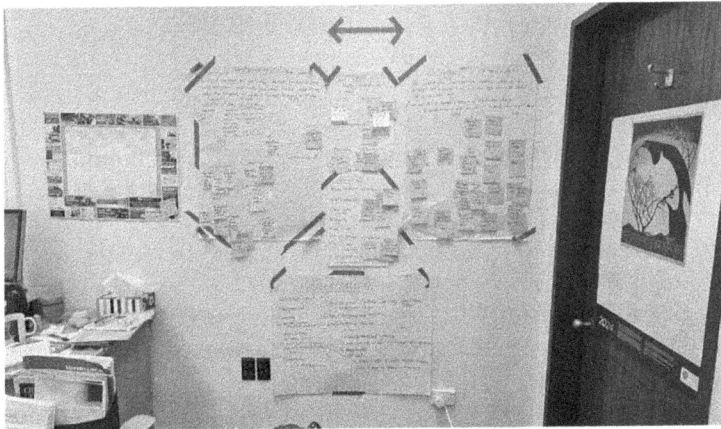

Figure 7.3 An example of this behind-the-scenes process of analysis from Anna Williams, a doctoral candidate who is using ethnography, autoethnography, and grounded theory in a study of Pākehā identity and well-being. The images show codes from her NVivo coding process that she printed and then arranged (and rearranged) around possible themes written on sticky notes, plus a wall-sized brainstorm of key theoretical concepts and ideas.

CREATIVE WAYS TO SHAKE THINGS LOOSE

Working continuously in only one narrow way is likely to generate more narrow and restrictive ways of thinking. Sometimes, the best, most original, interesting, or novel insights come from allowing yourself to see/think/feel through the data a little differently. Here are a few other creative activities that might help shake things loose a bit:

- **Talk it out:** Tell a non-expert about your research. Try to describe what you learned, what was interesting, or what felt new or surprising for them in everyday language.
- **Let pictures speak:** Choose three images to tell the story of your research. Why those three, individually, and what do they say *together*?
- **Director's chair:** Imagine you are going to turn this research into a play or a movie. Who would the main characters be? Where is it set? What genre would it be? What is the plot, what would be the climax, and how would it end?
- **Poetic vignettes:** Pick one particularly vivid moment or experience from your fieldwork and see if you can evoke how it sounded, smelt, and felt in your body in the form of a poem.
- **Remix:** Make a visual collage representing your field using a random magazine, or using copies of texts from your own fieldsite (e.g. media articles, brochures, or transcripts of your interviews). Think about what you are drawn to include and exclude, and how you are drawn to arranging each of the pieces in relation to each other on the page.
- **Sound of music:** What song or genre of music describes your fieldsite? Can you make a playlist that communicates something about it?
- **Art as insight:** What colour is your field? What shape or texture is your data? Find a medium that feels comfortable for you (e.g. crayons, pens, paint) and make a messy, abstract visual representation of your findings.

While some of these activities may help you deconstruct data, others may help you put it back into a meaningful whole again. Others still focus on unlocking the relational, sensuous, affective, imaginative,

Figure 7.4 An example of 'humuments' produced at a workshop run by UK-based sociologist Andrew Balmer, where participants were invited to layer their own meaning onto printed excerpts of interview data via painting (Balmer 2021).

playful, or poetic dimensions (Balmer 2021) and, thus, can bring to life things that you have begun to feel detached from throughout the analytic process. There are many more options too, so don't be afraid to experiment behind the scenes, where the stakes are low. The goal is to explore your data from all angles. Importantly, these sorts of processes can also be done *with* participants, in a playful way, at any stage of your research, or at multiple points throughout, as part of sense-making together.

ETHNOGRAPHIC THINKING, ETHNOGRAPHIC THEORISING

Ethnographers take care to identify and honour the specific places, times, settings, and relationships through which their findings were produced (McGranahan 2022). They know their work is situated. At the same time, most hope for their work to contribute to broader scholarly and public conversations. In order to do this, ethnographers have a few distinctive approaches to interpreting the wider significance of their findings, and bringing them into dialogue with theory. I focus on three thinking tools that are core to this.

CONTEXT, COMPARISON, AND CRITICAL THINKING

The specific observations recorded as data in your fieldnotes, photographs, interview transcripts, and archives, all took place in a context. Some of the immediate context may have been recorded in the data itself: for example, the social setting, place, or time. But they also occurred in a geographic context, historical context, political context, sociocultural context, economic context, ideological context, and so on. Ethnographers must continue to pay attention to these layers of context at *every* part of the process, including during analysis. The micro-level (personal and everyday) details observed provide clues about the macro-level (political, social, economic, and historical) structures, and vice versa.

STORY BOX: Comparing ballet cultures in three global cities

Helena Wulff studied three different national ballet companies in Stockholm, London, and New York in the mid-1990s. As a dancer herself, she conducted both participant-observation and interviews, including a process in which she accompanied dancers to performances by other dance companies. There have been debates in ethnographic literature about whether ballet is a 'global culture', given that it features specifically European fairy tales and folk legends and has a history embedded in nineteenth-century romanticism. Comparative ethnographic work, Wulff explained, "helps uncover the global connections" between different settings and styles of performance. Her work led her to argue that classical ballet is a "unitary form of physical culture" that continues to be produced and reproduced as it is transferred from body to body, in both training and performances (533). At the same time, she notes that ethnicity and nationalism are often negotiated through dance, "not least in colonial or postcolonial contexts", and her research was able to identify several contemporary examples in which productions had been adapted to wrestle with this in situationally specific ways (Wulff 2008).

Comparison is fundamental to how humans make sense of the world, but different research disciplines have different ways of using it as part of academic analysis (Felski and Friedman 2013). Ethnographers have a particularly interesting and, in some ways, particularly tricky relationship to comparison. On one hand, the mandate of ethnography, traditionally, is to provide focused, descriptive, holistic attention to a single case study at a time. This indicates an interest in depth rather than breadth. On the other hand, ethnography is about a committment to understanding things in a situated way, and comparison can help ethnographers identify how their findings are specific to those contexts.

Some ethnographers nominate **comparative** ethnography as a distinct methodological approach for their work. This means they design their research with a focus on gathering data in two or more sites as part of the same project, specifically in order to compare them. This is only one way of using comparison, in ethnographic work, however. For ethnographers focusing their research on just one field site, comparisons can always be made between their own data and the data that is provided by the existing **ethnographic record**, about similar phenomena in other places, or among other groups.

Comparison is built into the basis of ethnography in a more fundamental way, too, in that many ethnographers (depending on their positionality) are making comparisons between the people and places they are studying, and their own more familiar settings. This creates risks as well as opportunities (see also Chapter 4). Indeed, it is essential to recognise that comparison is never neutral but always develops within a history of hierarchical relations (Felski and Friedman 2013, p. 1), and it is easy for analysis to become laden with the ethnographer's own interpretive norms or values. Reckoning with these potentially **Eurocentric/Ethnocentric** tendencies has been a big part of the history of ethnography. Broadening inclusivity around who undertakes ethnographic research and from what positionality has been a helpful response, but so too has the lens of **cultural relativism**, which has been embedded in ethnographic practice from its earliest days. From the broader approach of decentering knowledge, "Comparison does not *automatically* authorise the perspective of those doing the comparing, but can also serve as a jolt to consciousness, initiating a destabilising, even humbling, awareness of the limitedness and contingency of one's own perspective" (Felski and Friedman 2013, p. 2, emphasis mine).

STORY BOX: Honouring the veil

Lila Abu-Lughod's study of honour and poetry in a Bedouin society, conducted in Western Egypt in the 1980s, became both famous and controversial. In particular, her analysis of veiling practices among the Awad 'Ali caused a strong response. In refusing to "condemn it as a sign of women's oppression", she was accused of being an apologist for a patriarchal society (1999, p. xx). When Abu-Lughod's book was reprinted fifteen years later, she was able to write a new preface addressing some of this and reflect on what her goals were in terms of representing the Arab world to Europeans from own her positionality as an ethnographer, and as the daughter of a Palestinian Arab. She wrote that

> One of my hopes was to foster great understanding and human sympathy. I thought this could be done through drawing a complex portrait of the society I felt I had come to understand. I wanted to explain the logic of the social system, including those aspects such as cousin marriage and women's veiling that are most abhorrent to outsiders.
>
> (xvi)

Recognising "the multiple meanings of veiling across the Muslim world today", she argued that Awad 'Ali veiling is "very particular" and was not necessarily akin to the 'new' type of piety veiling many readers were worried about. As such, she stood firm in not imposing on her participants the "critiques of morality that derive from foreign concepts" (xx), and instead trying to communicate how she had understood the practice being used by the people themselves, within their own social systems (Abu-Lughod 1999).

These concerns connect to something else that has had widespread value for ethnographic work: a critical lens. Critical analysis, in general, refers to steps taken to deconstruct and evaluate why things are the way they are, in the context of wider systems and institutions. The term '**critical theory**' more specifically refers to theoretical frameworks that focus on understanding the social world through a lens of power and power relations, often with a focus on facilitating social change. Drawing on both, some ethnographers anchor their work overtly in the methods, frames, and goals of **critical** ethnography, which shapes what they are looking for or interested in within their data, and what analytical frameworks they apply. But in other ways, a critical lens is woven into ethnographic thinking quite widely, power being something all ethnographers must meaningfully attend to in reflexive thinking about their own processes of research and representation, as well as in their way of deconstructing the fields they study.

ABSTRACTION AND GENERALISATION

Analysis is, in many ways, about abstraction. It is a process of working towards identifying general themes and concepts from situated and specific data. Abstraction is also the first step towards building theory. A theory is a way of making sense of the world. It is typically built of several interrelated concepts, offering "abstract propositions" about a group, society, or specific part of social life (O'Reilly 2012, p. 199). Yet for ethnographers, the particularities of social life observed in an ethnographic study are just that: *particular* to the place, setting, and people. This means ethnographers need to be very clear about the scope or **generalisability** of their findings – that is, is about their ability to draw broader conclusions from this one study. The actions and interactions in the field that produce our data can never be exactly reproduced. But while the details are not supposed to be taken as immediately representative of any larger group or population, they may be analysed to reveal *relationships* or *principles* that can be applied more broadly.

Because of this, ethnographic research offers powerful insights on key elements of social life such as power, identity, belonging, socialisation, and embodiment. Ethnographers contribute to theorising complex contemporary phenomenon such as bureaucratisation, migration, urbanisation, neoliberalisation, and globalisation. Ethnography typically does this not by proving or disproving any kind of universal laws but through its strength as a descriptive science. That is, it works by providing in-depth insights on how certain phenomena articulate within specific settings, and by using context, comparison, and critical thinking to think through what this reveals about human social life more broadly.

IN DIALOGUE WITH THEORY

Not all qualitative research has to lead to the development of theory, and some scholars have observed a reduced focus on theory following the **postmodern** turn (Daly 1997). Part of this may be that theories need some sort of logical consistency, but for ethnographers "it is important to realize that the nature of reality that is being theorized is complicated and filled with contradictions" (Daly 1997, p. 358). Ethnographers also have a different relationship to theory than do researchers in some other fields, with ethnographic research described as typically being "*in dialog* with theory rather than being *led or structured by* theory" (Pink 2009, p. 357, emphasis mine). Nonetheless, an ethnographer may choose to work with both established theory (also called *a priori* theory) or emergent theory. This choice is usually itself based on an **inductive** approach in which the approach to theory is determined through the process of analysis.

STORY BOX: Exchanging theories on value exchange

Many classical ethnographies of Melanesia explored and theorised how value was created through the exchange of both gifts and commodities (Orlandi 2017). Edda Cecilia Orlandi was not researching in Melanesia. Instead, she did her fieldwork in the warehouse of a supermarket in Milan, Italy. During this time she noticed that pallets were often exchanged between truckdrivers and warehousemen. Sitting on a pile of pallets herself, and scribbling in her notebook, the workers found her greatly amusing. Yet Orlandi was soon able to observe pallets as the subject of tricks, jokes, doubts, agreements, and exchanges and to learn about different types of pallets, how they were used, and why they mattered to the people who worked there. Applying some of the theoretical ideas that the work in Melanesia had established, around how reciprocity shapes social relations, Orlandi theorised that participating in pallet exchange was part of the creation of a community within the warehouse and between the different types of workers. As part of her analysis, she also compared her findings to another study that analysed the social purpose of corporate gifts in India. Orlandi used this develop and support her own conclusion that pallet exchange was a way for the workers to materialise their agencies and negotiate their status, thereby forming part of the ongoing production of social relations.

When applying existing theory, ethnographers may choose to work with one main theoretical framework, or they may take an 'eclectic' approach in which they connect to a variety of theories and theorists, as prove relevant (O'Reilly 2012, p. 199). Applying existing theories in a new way or to a new setting can lead to new insights; or again, to the ability to compare. Alternatively, ethnographers may develop their own theory or theories based on their analysis. This can look similar to or, in fact, draw directly from a grounded theory framework. Grounded theory comes from the work of Barney Glaser and Anselm Strauss in the 1960s and is used fairly widely throughout the social sciences today, including by ethnographers (O'Reilly 2012). It focuses on gathering **empirical** data, and working out what frameworks of understanding emerge directly from that. Grounded theory has key overlaps with an ethnographic approach, including encouraging researchers to draw on multiple forms of data, to reflexively draw on their own experiences, and to rely on an iterative and inductive process. Even so, ethnographers should be aware that the idea that a theory 'emerges' from the data can be a risky one.

WHOSE THEORY IS IT?

Theories are not really 'discovered'. Rather, they are made. As sociologist Kerry Daly put it, "Theory is, by its very nature, impositional" (Daly 1997, p. 35). As ethnographers organise and select data and construct explanations about this data, they layer on their own views. As such, Daly suggests that we can see theory as a type of story: a way that the researcher frames their own interpretations of social field or phenomena. "Like any stories, theories are products that are located in time and space, and they reflect the ideas, interests, and organization of the theory teller" (Daly 1997, p. 360). With this in mind, what makes a theory valid? The validity of ethnographic theory is often seen to come from its relationship to empirical data. But rather than leaning into the *authority* of theory, the process should stay embedded in reflexive discussions about the role of the researcher. This is not to invalidate it but to strengthen it, with the idea that theory is *most* generative when it invites scrutiny and critique (Daly 1997, p. 359).

It helps to recognise that the people we are studying almost always tell their own stories and make their own theories about the natural and social worlds they are living in. Anthropologist Sophie Chao discusses how the colonising mindset that is sometimes embedded in ethnography can mask this, especially for those working in marginalised or non-Western communities, since "theory is taken to be 'produced' by (and often for) the Global North, based on ethnographic realities that 'happen' in the Global South" (2022, p. 7). An alternative is to take seriously theory produced by participants at an intimate, everyday level. As Chao writes, "attending to theory in small places reveals the agentive and imaginative capacities of people in the face of structural inequalities that are relative to, but never totally determined by, macrolevel forces" (2022). Sometimes, an ethnographer's main task may simply be to work with participants to try and record these situated, unique, and everyday forms of sense-making, and communicate them to other audiences, but it is important to acknowledge if what they later publish and present is mostly a translation of vernacular theory. At other times, an ethnographer may be trying to put these **emic** knowledges into conversation with other **etic** frameworks or ideas. Many times, these tasks merge or overlap. Either way ethical reflexivity can be helped by exploring the tensions that may exist in the process through continuing to ask, as Daly puts, "whose theory is it?" (1997, p. 349).

PARTICIPANT FEEDBACK AND COLLABORATIVE SENSE-MAKING

Ways of knowing are always political. They have "moral trajectories with ethical directions and outcomes" (Daly 1997, p. 359). The process of theorising must, then, be mindful of the responsibilities of the ethnographer (see also Chapter 3) and the power dynamics involved in making claims about our participants social world. One key way an ethnographer can work out whether their theories actually fit with how their participants see and experience things, is by asking. This is sometimes called '**member checking**', a process that aligns with ethical obligations to check with how comfortable or happy people are with how we are representing their lives, their words, and their stories in anything we intend to publish or present. But this can go beyond the need to tick an ethical box. Instead it can become a measure of validity in itself, and a way to potentially improve and enrich our interpretations enormously by working on developing them *with* participants. This is best done iteratively, through layers of reflexive conversation at multiple stages of the project as part of "an ethics forged and continually renewed in and with community" (McGranahan 2022).

STORY BOX: The art of collaboration

Kryssi Staikidis is a US-based painter, teacher, and collaborative ethnographer. For more than twenty-five years, she has worked in relationship with Maya Tzu'tujil painting mentor Pedro Rafael González Chavajay in San Pedro La Laguna, Guatemala, and Maya Kaqchikel painting mentor Paula Nicho Cúmez in San Jan Comalapa, to develop Indigenous-centered understandings of visual arts education. Towards this goal, and inspired by Indigenous and **decolonising** methodologies, Staikidis has been mentored as a student by both Indigenous artists, who have guided the painting lessons. During such periods, Staikidis both lives and studies with each Maya mentor, who shares their cultural and community understandings through daily life experiences, dialogue, and art lessons emerging out of their home studios. The model Staikidis uses is collaborative ethnography, in which videotaped art lessons, conversations, and writing practices enable both teacher and student to ask questions and review the footage together (See Figure 7.5), so the artists themselves decide what segments are important and/or which to exclude (Staikidis 2020). Any films Staikidis edits with her mentors or publications she proposes are translated into Spanish and brought for review to participant mentors so that as their student, Staikidis can get their perspective, check for understanding, and seek approval for correct representations of their work and their words before presentation or publication.

The approach of collaborative ethnography can help embed this in a project from early stages. Of course, there can be complexities around when and how we show particular people our 'raw' data. This includes issues related to confidentiality and anonymisation. But waiting until the write-up is done and our interpretation locked in is still not the best approach. Rather, all ethnographers need to consider when and how they can fulfil the underlying ethos of ethnography as a situated, interpretive, and *relational* form of sense-making.

CHAPTER SUMMARY: KEY POINTS

- Ethnographic data often comes in multiple forms. Having 'enough' data relates to the richness of data and being able to triangulate between different forms of data to create a nuanced and holistic picture of the social field.

Figure 7.5 Kryssi Staikidis with her collaborator and mentor, Paula Nicho Cúmez, sitting in Paula's art studio to discuss some video data that Staikidis has begun to edit.

- Analysis of ethnographic data is an **iterative** and **interpretive** process, occurring over multiple stages, ideally in conversation with participants.
- Organising, categorising, and/or coding can support a 'focused revisiting' of the data in order to look for patterns or apply particular analytic frameworks. Writing about or visually representing the data can also help with seeing it in a new light.
- Ethnographers emphasise context, comparison, and critical analysis as key thinking tools, when interpreting their findings.
- Ethnographers may engage with existing theory as well as producing their own. They contribute to broader scholarly conversations about the social world, while aiming to recognise and honour the expertise of their participants in producing meaningful theories about their own lives.

RECOMMENDED FOR FURTHER READING

Daly, K. (1997). Re-placing theory in ethnography: a postmodern view. *Qualitative Inquiry*, 3(3), pp. 343–365. Available from: https://doi.org/10.1177/107780049700300306.

Denzin, N.K. (2017). *The Research Act: a theoretical introduction to sociological methods*. New York: Routledge. https://doi.org/10.4324/9781315134543.

Felski, R. and Friedman, S.S. (2013). *Comparison: theories, approaches, uses.* Baltimore: JHU Press.

McGranahan, C. (2022). Theory as ethics. *American Ethnologist*, 49, pp. 289–301. Available from: https://doi.org/10.1111/amet.13087.

Skjott Linneberg, M. and Korsgaard, S. (2019). Coding qualitative data: a synthesis guiding the novice. *Qualitative Research Journal*, 19(3), pp. 259–270. Available from: https://doi.org/10.1108/QRJ-12-2018-0012.

REFERENCES

Abu-Lughod, L. (1999). *Veiled sentiments: honor and poetry in a Bedouin Society*. Oakland: University of California Press.

Aldiabat, K.M. and Le Navenec, C.-L. (2018). Data saturation: the mysterious step in grounded theory methodology. *The Qualitative Report*, 23(1), pp. 245–261.

ARC. (no date). *Analysis & synthesis*. ARC. Available from: https://projectarc.design/process/analysis-and-synthesis [accessed 13 August 2024].

Balmer, A. (2021). Painting with data: alternative aesthetics of qualitative research. *The Sociological Review*, 69(6), pp. 1143–1161. Available from: https://doi.org/10.1177/0038026121991787.

Bönisch-Brednich, B. (2017). In praise of hunches. *Commoning Ethnography*, 1(1), pp. 152–158.

Chao, S. (2022). *In the shadow of the palms: more-than-human becomings in West Papua*. Durham: Duke University Press. Available from: https://doi.org/10.2307/j.ctv2j86bm4.

Daly, K. (1997). Re-placing theory in ethnography: a postmodern view. *Qualitative Inquiry*, 3(3), pp. 343–365. Available from: https://doi.org/10.1177/107780049700300306.

Davies, J.A. (2023). *Rongoā Māori: an autoethnographic account of my experiences with the Wairua*. Available from: https://hdl.handle.net/10292/15810 [accessed 26 January 2024].

Felski, R. and Friedman, S.S. (2013). *Comparison: theories, approaches, uses.* Baltimore: JHU Press.

Kleinman, S. and Copp, M.A. (1993). *Emotions and fieldwork*. London: SAGE Publications. Available from: https://doi.org/10.4135/9781412984041.

Landau, D. and Drori, I. (2008). Narratives as sensemaking accounts: the case of an R&D laboratory. *Journal of Organizational Change Management*, 21(6), pp. 701–720. Available from: https://doi.org/10.1108/09534810810915736.

McGranahan, C. (2022). Theory as ethics. *American Ethnologist*, 49(3), pp. 289–301. Available from: https://doi.org/10.1111/amet.13087.

O'Reilly, K. (2012). *Ethnographic methods*. Florence: Taylor & Francis Group. Available from: http://ebookcentral.proquest.com/lib/otago/detail.action?docID=958470 [accessed 28 September 2022].

Orlandi, E.C. (2017). The values of pallets: an ethnography of exchange in the warehouse of an Italian supermarket. *Journal of Material Culture*, 22(1), pp. 19–33. Available from: https://doi.org/10.1177/1359183516662674.

Pink, S. (2015). *Doing sensory ethnography*. SAGE Publications Ltd. https://doi.org/10.4135/9781473917057.

Skjott Linneberg, M. and Korsgaard, S. (2019). Coding qualitative data: a synthesis guiding the novice. *Qualitative Research Journal*, 19(3), pp. 259–270. Available from: https://doi.org/10.1108/QRJ-12-2018-0012.

Staikidis, K. (2020). *Artistic mentoring as a decolonizing methodology*. Brill. Available from: https://brill.com/display/title/54333 [accessed 14 August 2024].

Strathern, M. (2004). *Commons and borderlands: working papers on interdisciplinarity, accountability and the flow of knowledge*. Syracuse: Sean Kingston Pub.

Wulff, H. (2008). Ethereal expression: paradoxes of ballet as a global physical culture. *Ethnography*, 9(4), pp. 518–535.

8

COMMUNICATING ETHNOGRAPHIC KNOWLEDGE, CREATING ETHNOGRAPHIC TEXTS

Communicating knowledge is a key part of the ethnographic endeavour. The fact that the term 'ethnography' describes both a research methodology and a genre of text is telling in itself. This final chapter deals with the outward face of ethnography – what to do with findings when you have them. It emphasises that communication should not be just an afterthought in the research process but something considered right from the start, showing how decisions about the form in which ethnographic data is communicated connect back to central questions around the purpose and significance of ethnography. A 'text' here is taken to be not only a written form but any type of communication artifact, including the visual, audiovisual, embodied, digital, and multimodal. Ethnographers, in fact, draw from a variety of these traditions to present diverse and engaging texts to the world. This chapter discusses how each offers different affordances for responding to the challenges, opportunities, and ethical responsibilities involved in ethnographic communication, including the potential for collaborative *co*-construction of texts. It opens up some of the key tensions ethnographers face in responding to debates around **objectivity**, **subjectivity**, authority, authorship, voice, and truth. The final section considers the impact of our work and the way we might think strategically about engagement with a variety of different audiences and publics. How do we make what we do useful, meaningful, and accessible?

DOI: 10.4324/9781003404880-8

THE ROLE OF ETHNOGRAPHIC TEXTS IN ETHNOGRAPHIC RESEARCH

Producing texts is part of the process of producing (and reproducing) ethnographic knowledge. The question of 'how do we know what we know?' is intimately tied with the question of 'How do we communicate what we know?' For this reason, ethnography is not 'done' until it takes some type of textual form. Crafting an ethnographic text isn't about communicating *all* your data or mapping out your **field** in its entirety, however, but rather about selecting a particular idea to share, argument to make, or story to tell (O'Reilly 2012). This also involves picking a form, medium, and genre to tell it in, and using particular strategies and techniques to do so effectively. The decisions about how to represent what you have found are also part of the sense-making process, especially since the process doesn't happen as a one-off but through many layers of decisions – as part of editing, revisiting data, reworking structure, coming up with headings or captions, and so on – whilst also being a deeply collaborative practice. This process reveals something important: that the ethnographer isn't just a neutral recorder but rather a translator, interpreter, and storyteller.

The ethnographer's investment in an identity as 'storyteller' has held constant through many decades of change to ethnographic practice (McGranahan 2015). In many ways, as Carole McGranahan emphasises, it is this that best highlights the complex relational and ethical contexts in which "We tell stories to make points; we join others in growing worlds from stories; we are trusted with people's stories and commit to tell them the right way to new audiences" (McGranahan 2022). With all this in mind, ethnographers can find the task of deciding how to share their findings to be challenging not only technically but emotionally. This is understandable: there are stakes for the audiences, the **participants**, and the ethnographer themselves. Yet there is a vast record of meaningful and impactful ethnographic texts, across many genres, to inspire us to the task: texts that contribute to a vivid and empathetic record of the challenges and joys of human life, texts that challenge taken-for-granted assumptions about the world, texts that call for change where it is needed, texts that tell stories that wouldn't otherwise be told. This makes the task of grappling with the 'how' well worthwhile.

WHAT MAKES A TEXT ETHNOGRAPHIC?

Ethnographic texts are diverse in focus and form. They may also seem, at times, to resemble other genres of writing or communication – from documentaries to travel stories to autobiographies – and indeed, they can employ similar communicative strategies. So what makes an ethnographic text *ethnographic*? While over time, "The nature of ethnographic evidence, interpretation, authority, style may indeed have changed" (Van Maanen 2006, p. 16), there remain some common features or qualities that work to align the text with the goals of an **ethnographic lens** (see Chapter 1).

Firstly, ethnographic texts are descriptive and evocative. They work to create a detailed and textured portrait of specific people in a specific place and time. The ability to do this comes from methods of gathering and recording data which emphasise insider perspecitves and experiential knowledge gained from 'being there' (see Chapter 4). Many ethnographers have invested in honing literary techniques that assist in creating a *sense* of being there for the audience, inviting them into a specific moment in time, a specific place, a specific set of social relations. Ethnographic texts also aim to place what is described *in context.* This is facilitated by **thick description** (see also Chapter 5), which enables them to not only describe but to interpret and analyse. In this sense, ethnography is just storytelling, but a form of "*theoretical* storytelling" (Shah 2022, p. 28, emphasis mine). Ethnographic texts are also sometimes defined by their relationship with 'the field' and/or some kind of **fieldwork**. In other words they are expected to be informed fairly directly by data gathered in and about real-world settings, or through talking and collaborating with actual people. It is interesting to know, however, that some texts that have been described as 'ethnographic' have been produced by people with little or no formal training in ethnographic methods but who have still been able to embody these qualities in compelling ways.

TYPES OF ETHNOGRAPHIC TEXT

Whilst the ethnographic lens has acted as a sort of centrifugal force, ways of representing ethnographic knowledge have also shifted considerably over time. Some of these changes have been driven by technological advances. Others have been shaped by debates around the nature and ethics of ethnographic knowledge-making. Today, ethnographers are extremely lucky in that we have our choice of many different mediums, forms, and styles of text with which to communicate. Some paths are more well-trodden than others, and some are more recognised and valued by mainstream academic or research institutions. Nonetheless, there are examples of ethnographers working in almost every style and medium, and a general willingness to experiment thoughtfully with new forms too. This includes (as a nonexhaustive list, and with several overlapping categories): *ethnographic novels, ethnographic fiction, ethnographic poetry, flash ethnography, visual ethnography, graphic ethnography, ethnographic film, ethnographic photography, sounded ethnography, ethnographic art, ethnographic performance/theatre, ethnographic dance, multimodal ethnographic, experimental ethnography.*

An ethnographer's choice of textual form for sharing their work should consider the purpose and the audience. Ruth Behar advocates for exposing ourselves to a wide range of textual mediums as readers and audiences, to see what we might learn from each (2020). At the same time the choice may relate to the individual researcher's skills and interests, and can develop into its own form of research praxis through being enmeshed with processes of recording or questioning in earlier phases too (see Chapters 5 and 6). Each genre or medium of communication has a different history and different affordances. Since this chapter doesn't have space for a *detailed* overview of each, it focuses instead on discussing, within broader categories, how different mediums have helped ethnographers respond to bigger practical, ethical, or **epistemological** challenges.

ETHNOGRAPHERS WRITING ETHNOGRAPHY

"What do ethnographers do?" Clifford Geertz once asked. "They write!" (Geertz and Darnton 1973, p. 19). Ethnography has long been considered a discipline of writing. Ethnographers are expected to engage with writing all stages of the research project, from writing plans or proposals to writing **fieldnotes** to describing findings. This is because writing is seen as part of thinking – part of the process of sense-making, analysis, and **reflexivity** – not just the product of it. It is difficult to standardise the method of ethnographic writing, however (Stinnett 2012). Ethnographic writing can take on a diverse range of different styles. Ethnographers write journal articles and book chapters. They may write media pieces or public blogs, as I return to discussing later. They also produce their own monographis, to offer a detailed particular of a particular field, group, or social practice, which is most often what people refer to when they refer to a text as 'an ethnography'. Ethnographers have produced many unique books that have been engaging to public as well as academic audiences, some becoming quite famous or influential: from Margaret Mead's *Coming of Age in Samoa* (1928) to Ruth Benedict's *The Chrysanthemum and the Sword* (1946) to Marjorie Shostak's *Nisa: The Life and Words of a !Kung Woman* (1981) to Jeff Shonberg and Philippe Bourgois's *Righteous Dopefiend* (2009) and Anna Tsing's *The Mushroom at the End of the World* (2015).

Many ethnographers have also embraced literary styles of writing. For some ethnographers, this has meant drawing on creative, narrative, lyrical, or poetic techniques *within* more academic texts, but for others, it has led to working in genres such as the novel, the creative essay, or in short forms such as poetry, short story, or flash fiction. In all cases, creative writing can help with evoking a field in memorable and engaging ways but, at the same time, can provide alternative and more humanistic way to explore theoretical ideas, at the level of the body and the senses, temporality, and relationality.

STORY BOX: Poetry and violence

Nomi Stone is a poet and an ethnographer who carried out fieldwork at the military training sites around the United States where American soldiers were playing out simulated conflict scenarios in mock villages and cities to prepare for deployment in Iraq. At these sites, former Iraqi interpreters and refugees took on roles as actors, bringing the training scenarios to life. Stone has published two collections of ethnographic poetry about this. The key question for her is "how to render the condition of being in a body and being in time, how to represent the textures of living in the world: its crisscrossing structures, its constrictions and openings" (Stone 2020). Poems, she suggests, provide their own secret tools, working as an "embodied laboratory, to amplify sensation, to make a lived world". This is evident in a multipart poem called 'Drones: an exercise in awe-terror", which is based on fieldwork at Creech Air Force Base in Nevada and invites us to see (and feel) through the eyes of a drone pilot, where we begin with

A sea of, a drowning of–everything seems
to be red rock. Prickling of dust and salt.
Seething, the sun between
the shrubs.

Overall, Stone suggests that poetry's work is to bring the "violent ghosts" of things like war to consciousness by equipping ethnographers to "open language and shake it loose" so we might "feel the sharp aroma of words on the page reach our bodies" (BOMB Magazine).

History turns, literary turns

The **interpretive** turn, the critical turn, the reflexive turn, and the **postmodern** turn, all challenged established representational norms in ethnographic writing. Also key was the 'literary turn' of the 1980s, a time in which ethnographic texts began to be recognised and analysed as *texts,* with an eye to deconstructing the literary techniques they commonly used, for example, narrative structure, characterisation techniques, plot lines, authorial voices, imagery, phrasing, allusion, analogy, and more (Van Maanen 2010). A major contributor to this literary turn, American sociologist John van Maanen famously identified several distinct genres of ethnographic 'tale'. The first was the realist tale, referring to ethnographic texts that adopted a formal, institutional voice to lay out some 'objective truth' about the people, culture, or social field being discussed, in an authoritative way. The second was the confessional tale, which focused on the researcher's own presence in the field through presenting, the story of their (sometimes stumbling) experiences of fieldwork. The third, the impressionist tale, is described as emphasising vivid, startling, or emotionally charged ways of evoking the field and drawing the reader into it. These 'tales' became a useful way to think through some of the bigger questions of ethnographic representation, though in actuality, they often blend and have also since "fragmented into several emerging styles" (Van Maanen 2010, p. 246).

Literary techniques are closely entangled with the politics of knowledge. Another example of this was the critiques of 'the ethnographic present'. This refers to a trope of ethnographic writing in which ethnographers describe their field in the present tense. This can lend a sort of (false) authority to ethnographic accounts, implying that culture is something static and that the researcher has uncovered some objective, unchanging truth, while masking the very specific contexts, moments, and encounters through which the account was produced. In response, ethnographers today mostly avoid using the ethnographic present in their writing, and instead have focused on "locating their ethnographies historically, spatially and structurally", including in relations of power, time, politics, and technological developments (O'Reilly 2012, p. 213). This is equally a response to debates about 'writing **culture**' in the 1990s that continued to push ethnographers to "find ways to write that work against the typification of communities that made them into distinct and alien cultures" and to recognise the relations of power that fuelled this (Abu-Lughod 2000, p. 262). These changes have come together to help ethnographers lean into writing that contributes to critical, decentered, and **decolonising** work.

Over time, the task of ethnographers – to represent culture – has "become heavier, messier, and less easily located in time or space" (Van Maanen 2010, p. 245). The postmodern turn has meant that much of the work in recent eras has been to "accept and celebrate the complex, ambiguous, messy nature of the social world and ethnographic research" through efforts to move away from the need to provide "neat, ordered narrative accounts written in an authoritative voice" (O'Reilly 2012, p. 257). This involves many different possibilities for experimenting with fragments or with multiple voices, allowing for complexity and pointing to multiplicities of ways of living and being human.

STORY BOX: The text multiple

Medical anthropologist Anne-Marie Mol conducted two years of fieldwork at a large Dutch hospital in order to study a disease called atherosclerosis. What fascinated her were the multiple ways this disease was brought into being as a social artefact by and for both patients and clinical staff, generating paradoxes and multiplicities. The book she published from this work is called *The Body Multiple: Ontology in Medical Practice* (2003). This book experiments with the structure and form of an academic text, with the main chapters taking an unusual form: splitting each page horizontally into two different parts. The first part (the 'upper text') provides an ethnographic account of the hospital, including thick description and lengthy italicised quotes from Mol's interviews. In the second part (the 'subtext'), which appears in a different font and visual format, Mol discusses the first part and creates a reflexive dialogue between this and other scholarly literature. Her approach interrupts the normal ways of engaging with a text, and her preface notes that people may have to "invent a way of reading that works for them" to proceed through it (Mol 2003, p. ix).

STORY BOX: Unfolding fortunes in multiple languages

Yixuan Wang used digital ethnography to study the experiences of U.S. teachers involved in China's English-language teaching industry, exploring themes of privilege, marginalisation, precarity, intersectional identities, and the bilingual/bicultural. Wang started with an analysis of 100 job advertisements from five different social media groups. She initially worked on writing poem in more traditional stanzas, to express some of the themes she established from **qualitative coding** of this material. But as interviews with participants revealed more complex and intertwined experiences of commodification, fetishisation, discrimination, and scamming, Wang decided to break away from linearity. The final version of the poem, which won second place in the 2023 Society for Humanistic Anthropology Ethnographic Poetry Prize, uses an experimental form which can be printed and turned into the schoolyard game which is sometimes referred to as a 'fortune teller' or a 'cootie catcher' (see Figure 8.1), and which gives readers multiple and self-guided ways to engage, aiming to "provide a *multimodal* and multidimensional avenue for the audience to become co-investigators to appreciate and experience participants' understandings" (Wang 2024/forthcoming).

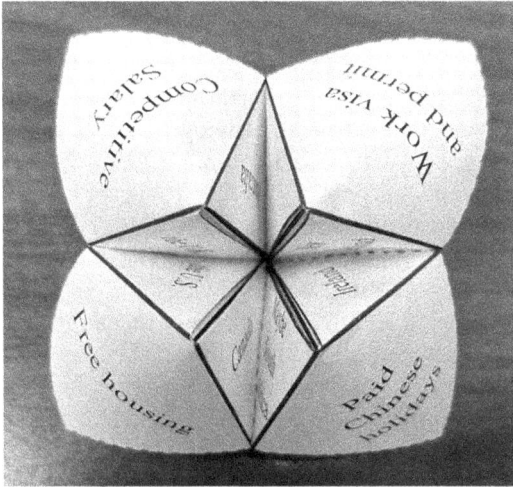

A Cootie Catcher to Recruit "English Native Teachers" to Teach in China

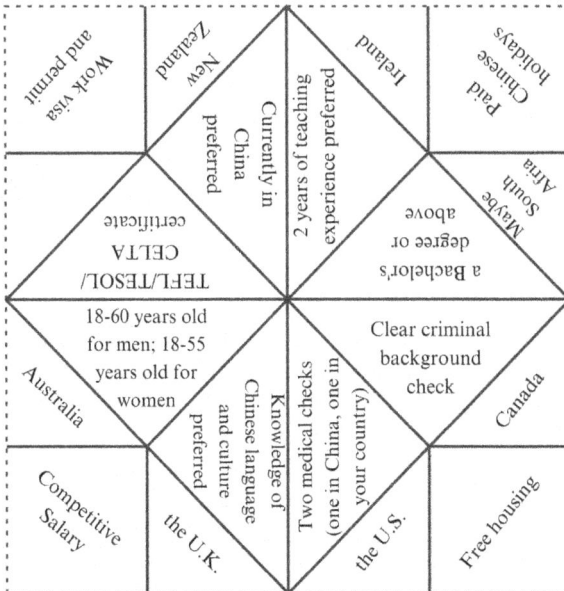

Figure 8.1 Top: the poem as printed on the page. Bottom: the poem as assembled into its 'fortune teller' form.

VISUALISING ETHNOGRAPHY

If ethnography is a 'way of seeing' (Wolcott 2008), then visual mediums are surely a fitting form of communication for ethnographic knowledge. It is true that ethnography has a long and rich history of engaging with visual communication in terms of both recording data and representing findings. As charted in Chapter 5, ethnographers have often been at the cutting edge of developments in the use of visual, media, and digital technologies, including photography and film. Indeed the camera created new ways for ethnographers to communicate a sense of 'being there'. There can be many paradoxes in the production of ethnographic 'truth' through visual mediums, however. For example, in Robert J. Flaherty's famous film *Nanook of the North* (1922), which attempted to realistically portray the lives of Inuit people, the sense that many Western audiences had of seeing intimately into a different cultural world was created through careful staging and the use of actors. Ethnographic texts are expected to deal reflexively with their own processes of interpretation and representation. Yet just like written texts, visual texts can easily exclude or 'edit out' the person and the processes occurring *behind* the camera, so care must be taken to recognise how they are selective, perspectival, and constructed, emerging from specific choices by ethnographers/filmmakers, as well as through collaboration and negotiation with participants.

Ethnographers have also often promoted visual texts as a particularly suitable way to engage wider audiences. Margaret Mead, for example, used ethnographic film in a very public-facing manner. But to what extent are images stand-alone versus requiring some other form of (written or verbal) text to frame them? While Mead valued the ability of film to include the participants' own voices, she also tended towards a strategy of combining film footage with intertitles and didactive voice-overs that interpreted the material (Alfonso et al. 2004). This hints at points of suspicion and anxiety that have sometimes arisen around the capacity of visual images to communicate (Chio 2021, p. 4). *Does the audience have the tools they need to interpret what they are seeing? Could it be misread?* Ethnographers cannot always control the meanings audiences take from a visual text. An example of the

Figure 8.2 A series of social media–inspired photographic self-portraits taken by young Somali refugees in Delhi, India, being presented here in a gallery as part of a 'talk-back' session organised by ethnographer Ethiraj Gabriel Dattatreyan in 2014. (Dattatreyan 2015).

ambiguous potential of representing cultural practices on film is *Les Maîtres Fous*, a famous and controversial ethnographic film produced by French film director Jean Rouche, released in 1957, and subsequently banned in several places. The film focused on a ritual associated with the African 'hauka' movement in Nigeria, but the audience was divided over it, with some people arguing that it entrenched racist and imperialist views of 'savage' Africans, and others interpreting it (and the ritual it portrayed) as subversively anticolonial (Meyer 2017).

There has been a wealth of **critical** scholarly attention to the relationship between visual technologies and violent or oppressive structures of power (Chio 2021). This has included attention to both the colonial gaze and the ethnographic gaze. At the same time, visual mediums can open up different potentials for **collaborative** processes of making texts. Through these, the power relations of the gaze can also be reconfigured, as individuals and communities might also find ways to *self*-represent (see also Chapter 5), or to work actively and co-constructively with ethnographers.

Figure 8.3 Illustrations from the 'Reclaiming Dita' project by Boris Stapić.

Meanwhile, there remain many ways of creating visual texts without a camera. Drawing has been part of the ethnographer's toolkit from early day. Line-making, as Ingold expresses it, is a powerful and fundamental way of engaging with the world in which movement, observation, and description "become one" (Ingold 2011). Drawing embeds references to the body of the creator – their line of sight, the movement of their hand – in its very form. There is no mistaking drawings or illustrations for any kind of authoritative 'truth', but rather, they represent a wholehearted leaning into subjectivity. This has a vast range of potential applications for ethnographers.

A recent resurgence of interest in 'graphic ethnography' has highlighted new ways this might feed into accessible forms of text. Sequential art – which can be either long form or short form and can range in style from roughly sketched comics to highly rendered illustrations – holds inventive possibilities in its ability to draw on, combine, and remix the conventions of other mediums (for example, newspapers, graffiti, or photography) (Gilbert and Kurtović 2022). It potentially offers a different way of relating to existing texts and communication genres from the relevant field. It also holds some unique possibilities for storytelling, particularly in its way of "bringing different **temporalities** into relationship" and helping to "visualize and materialize the immaterial and overlooked aspects of politics" (Gilbert and Kurtović 2022).

STORY BOX: Producing detergent, producing change

On the outskirts of the post-industrial city of Tuzla, in Bosnia, is a detergent factory called 'Dita'. Andrew Gilbert and Larisa Kurtović had been involved in ethnographic research with the workers of Dita and their local allies for several years, documenting how the union mobilised, and finally won an unprecedented victory, preserving the factory and restarting production despite the threat of bankruptcy in 2014. In 2018, looking for new ways of working collaboratively with industrial workers, activists, artists, and politically engaged local academics, Gilbert and Kurtović teamed up with Sarajevo-based graphic artist Boris Stapić. Producing graphic work helped them "identify and render visible by rendering visual critical elements of the workers' struggle that are often overlooked, dispersed in space and time, and considered uneventful". The Reclaiming Dita project was thus their activist-based response to some of the more "extractive" tendencies of their discipline, towards the possibility of a text that might "seed the political imagination of readers" and inspire other activism in the region and beyond.

Ethnographers may produce artworks in any number of visual and multimodal mediums, including painting, sculpture, or installations. All are potentially rich ways of facilitating public engagement and enlivening new perspectives on a topic or field. Ethnographic art can be shared as part of art exhibitions or installations, or reproduced in books, journal articles, or other digital and print mediums. At times, ethnographers may focus these activities on sharing their own creations or sharing co-creations from the research process, while at other times they may prioritise scaffolding a space for participants' own or existing creative work to be recognised, valued, and shared in different settings or with new audiences.

Multimodal and experimental ethnography

All ethnographic communication involves a translation of life as lived into one or more mediums of textual communication. The

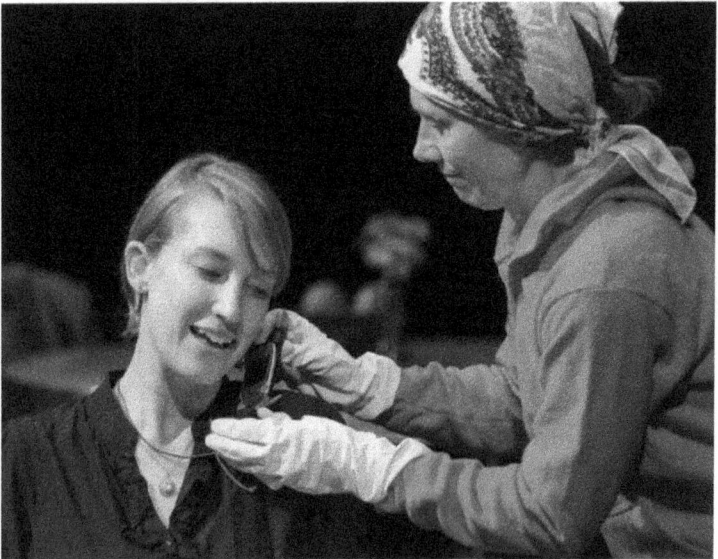

Figure 8.4 Photo from a staging of *I Was Never Alone*.
Source: Photo credit: Jim Carmody.

pressing question is how not to lose the richness and wholeness of lived experiences, and of social worlds, in this process (see also Chapter 5). Ethnographers have responded to this challenge in a variety of ways. One response is to produce texts in embodied forms using various genres of performance, including, for example, through theatre or dance. Performance 'texts' may be derived in various ways from **empirical** ethnographic material. This may include collaborative processes through which participants take an active role in scripting or performing their own stories.

> **STORY BOX: Staging stories**
>
> Cassandra Hartblay conducted ethnographic fieldwork in Russia in 2012–2012. Three years later, she staged an ethnographic play based on this work. *I Was Never Alone* is a ninety-minute documentary play featuring a series of monologues (or 'portraits') "composed of quotes transcribed from ethnographic interviews with real people, whose life experiences form the inspiration for each character" (Hartblay 2024). The tone of these monologues ranges from poignant and political to humorous and everyday as they deal with issues around the segregation of people with disabilities as well as broader themes of love, family, alienation, and the desire for connection in a digital world. It was staged and performed multiple times in the USA from 2015–2018.

The idea of multimodal ethnography has been used to reflect changing and diverse technologies and practices of communication, i.e. things that can't be neatly boxed as one single mode. A rising focus on multimodal work responds to some of the epistemological shifts mentioned earlier, emphasising and valuing a multiplicity of ways of knowing and sharing ethnographic knowledge. It also recognises that many types of text are already communicating in multiple modes, so for example while theatre is a multimodal medium, in another sense so is a print photograph, since it is both visual and material. Digital tools and technologies are typically multimodal as well.

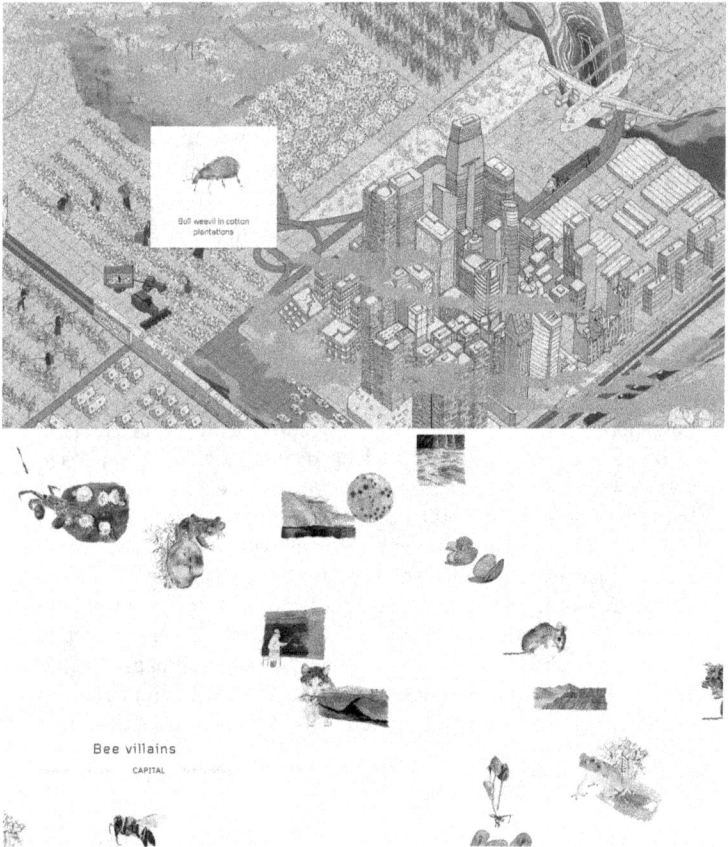

Figure 8.5 Screenshots from the Feral Atlas website.

Source: Taken February 2025, from https://feralatlas.supdigital.org/.

The term 'experimental' can be a way of describing ethnographic work which uses new technologies, uses existing technologies for new purposes, or finds a way to subvert established or traditional uses of a particular medium, genre or form, which can include experimental writing too.

STORY BOX: Feral digital communication

Anna Tsing works in the area of the multispecies and more-than-human. Her book *The Mushroom at the End of the World* (2015), has been described as both speculative and experimental, using a multisited and multisensorial approach to exploring human transformations of the planet. Yet Tsing followed this up with something even more experimental: a collaborative, curatorial digital project called *The Feral Atlas* (2021). What appears first on this website is an expansive hand-illustrated map, resolving into a series of interactive elements, where users can click on different entities 'drifting' across the screen (see Figure 8.5). In each of the pages they lead to, users find video and audio poems, which the creators state as involving pleasures that themselves can "grab readers' sensibilities and twist them in unexpected ways" (Deger et al. 2021). These are used as the gateway into longer "field reports" based on ethnographic case studies, each presenting specific **situated** examples of "feral ecologies".

CHALLENGES AND OPPORTUNITIES IN ETHNOGRAPHIC COMMUNICATION

As is hopefully becoming clear, the process of communicating ethnographic knowledge is inextricably tied with the epistemological foundations of our work, which is, in turn, tied to ethical and representational questions. The following section distils some of the tensions that suffuse the decisions ethnographers have to make.

OBJECTIVITY, SUBJECTIVITY, AND TRUTH

As Murphy and Dingwall write, our assumptions about the nature of reality, the possible knowledge of that reality, and the status of truth claims all have implications for our judgements about the ethnographer's responsibilities (2001, p. 2), including our responsibilities to represent ethnographic 'truth'. Ethnography has historically involved a "blurring of genres" between humanistic and scientific modes of inquiry (Daly 1997). Different ways of writing ethnography have leaned on objective or subjective voices in different ways, some seeming to claim a more direct correlation to truth based on their own (typically 'scientific') stylistic norms. Critical, feminist, poststructuralist, and postcolonial schools of thought have successfully questioned and deconstructed ideas of the 'naturalness' of ethnographic description by highlighting the constructed, interpretive, positioned nature of all ethnographic accounts. This led to a rejection of the idea of a 'view from nowhere' in favour of recognising that ethnography is always a 'view from somewhere'. An embrace of subjectivity and the use of a first-person narrative account has been favoured by many ethnographers as a way of responding to this, and a strategy for achieving an explicit presence in the text. Yet a focus on empirical data remains, meaning ethnographies can often come across as a strange cross between author-saturated and author-evacuated text: "neither romance not lab report, but something in between" (Geertz, in Behar 1996). Stating one's positionality and presence is only the first step. As Ruth Behar emphasises, the scholar must also "Draw deeper connections between one's personal experience and the subject under study" (1996, p. 13). In **critical** ethnography, this more developed form of self-reflexivity is the single strongest strategy for combatting traditional **positivist** and authoritative approaches (Stinnett 2012).

At the same time, "A good part of our writing is both explicitly and implicitly designed to persuade others that we know what we are talking about and they ought therefore to pay attention to what we are saying", as writes Van Maanen (2010, p. 240). Ethnographers have grappled with how to be appropriately reflexive while still establishing the legitimacy of what they present. For a long time, "experience" was seen as key grounds for ethnographic authority or for the legitimacy of an ethnographic text. The culmination of this was that some classical ethnographies were critiqued as focusing more on conveying the ethnographer's personal knowledge of the field than communicating the voices or viewpoints of their interlocutors (Pandian 2019, p. 48).

POSITIONALITY, VOICE, AND POLYVOCALITY

In ethnographic texts, we expect to see a focus on emic understandings and insider perspectives. How, then, are these represented through the voice of the ethnographer as the assumed author or creator of these texts? The answer depends partly on the ethnographer's

positionality, of course, which can itself be complex, multistranded, or shifting (see also Chapter 3). *To what degree (and if so, on what basis) do you claim insider knowledge?* Additionally, it can be the communication style or technique which contributes to a text taking on a particular voice. This quickly links to the issues of representational ethics; is the text then speaking *about* or *for* the people it represents? This has led many ethnographers to think about how knowledge can instead be included in the participants' own voices. While extended quotes are common in ethnographic texts and can help with this, there are more possibilities available, including the development of multivocal or polyvocal texts, that is, texts in which participants not only are quoted but are active narrators and co-constructors. Ghassan Hage, writing to to explore an ethical response to the racist texts embedded all throughout in the histories of academia, urges ethnographers to avoid placing participants as a 'passive subject', and instead think about ways of writing with them (Hage 2020).

The term 'dialogical ethnography' has sometimes between used to refer to texts which present "a dialogue between the researcher and the researched" (O'Reilly 2012, p. 218). As O'Reilly explains, this can be a strategy for "reflecting the difficulties of interpreting another person's world, the dialogic nature of research itself, and the involved, messy, subjective and emotional nature of ethnographic research" (ibid). Collaborative and participatory modes of research and writing lend themselves much more to this but may present their own dilemmas, especially when pursued in institutional contexts that are entrenched in Western scientific/historical norms: *Who will be credited as an author? What counts as 'expert' knowledge?*

Autoethnographic work can be another powerful way of making the writer both the subject and the object of the story – positioning them as an expert based on both lived experience and academic analysis – allowing them to make interpretations of their own experiences in their own words. This has been taken as particularly appropriate and useful for those speaking from marginalised positions. It has been recognised as form of text in which Indigenous researchers can voice aspects of their own culture and history in their own voice (White 2010). Autoethnography can have its own challenges, however, as it rarely speaks *only* about the author and, therefore, still takes on the task of representing others through the author's frame of reference.

FICTION, INVENTION, AND IMAGINATION

Ethnographers are committed to reflexive representations of real people, places, and events. Because of this, the idea of 'ethnographic fiction' has caused some perplexity. Part of this uncertainty comes from acknowledging that what counts as 'fiction' can be fuzzier than it first appears. As Falcone points out, all ethnographers write somewhere on a continuum 'factual' data and fictional or fictionalised representations (2020). While some work self-consciously in genres such as the ethnographic novel, most must use a degree of creative license *within* more standard ethnographic texts, to translate and simplify complex social wholes into tangible, accessible stories. Of course, there *are* distinctions between 'making' and 'making up' texts (Falcone 2020) and presenting wholly fictional scenarios in a deceptive way (i.e. without disclosing the process) would be unethical. But applied reflexively and transparently, fiction can hold rich potential for ethnographers, extending rather than breaking from existing methods through drawing on empirical data to inform vivid world-building, often in order to point out taken-for-granted aspects of a given world (Watson 2022).

To qualify as ethnographic, fiction should show convincing understanding of the actual forces and relationships that animate the worlds it depicts, to the point that it feels it presents *could* have happened. . . even if it didn't. Of course, some fiction stretches beyond existing settings, and there is also a strong historical association between anthropological, sociological, and ethnographic work and speculative genres such as science fiction or dystopian fiction. Rather than trying to mirror reality, ethnographers working in these spaces seek to explore possibilities and stir reflexive thinking in a way that remains connected to the intimate particularities of a place, community, or practice. Ultimately, when fiction is clear about its purpose – that is, in proposing open-ended questions about the social world rather than making authoritative truth claims – it can become a great tool for ethnographers.

AUDIENCE AND IMPACT

Why do I write? What is our purpose? Who is our reader? How do we navigate the different tensions we face – the constraints of academic evaluation criteria versus the compulsions of writing for wider publics, scholarly fidelity versus activist commitments, writing as anthropologists versus producing journalism or fiction? (Shah 2022, p. 2)

We create ethnographic texts in hopes they will have an impact on the world. But on whom, specifically? And what *sort* of impact? The questions Alpa Shah poses are valid not only for ethnographic writing but for all forms of ethnographic communication.

STORY BOX: Radical writing

In 2010, Alpa Shah, a British anthropologist, went on a 250-kilometer night march with members of a Naxalite guerilla army in India. When she returned to her office in London, she grappled with how to write up her data from this extraordinary experience, and from other parts of her field-work in India. As she wrestled with this, and considered the idea of moving outside of more 'traditional' academic writing modes in response, she was warned against taking up the "creative and experimental" path by some other academics. Yet eventually she committed to a public-facing book. The process of learning to write for the public, for Shah, meant unlearning habits of academic writing. "There is no blueprint, no model, no prefigured ideal", she affirmed. Instead, she tried to learn from writers of fiction, aiming to write without jargon in an accessible and engaging way, even while she distinguished the text she was producing from either fiction or journalism. Her book, *Nightmarch: Among the Guerrillas of India's Revolutionary Movement*, was published in 2018 and has since won several awards. The project helped Shah "reclaim the radical insights offered by our ethnographic research", including their potential to create the type of knowledge that challenges existing hegemonic structures and norms, which is a powerful thing to do in the public eye.

REPRESENTATION, POWER, AND SUFFERING

Questions of power are present in all forms of representational prac-tice. *Who has the right to represent whom? Who is the ethnogra-pher beholden to, what are their goals, and what is at stake?* Many of the people that ethnographers work with are capable of represent-ing themselves and will continue to do so, regardless of any sort of representations an ethnographer generates. However, others, for myriad cultural, social, political, and personal reasons, are denied the opportunity to narrate their own lives (McGranahan 2020). Ethnographers are responsible for the representations of people, communities, and practices that they put into the world. This means they are accountable for inaccuracies and errors. But it goes further than this. They are also accountable for representations that, while not strictly false, may, through partial or positioned representations, produce or reproduce stereotypes or result in **othering** or dehumanisation. This can lead to many sorts of harms, including direct interpersonal harms (of discrimination or violence) or the reproduction of broader systemic discrimination and structural violence. Special care must be taken when dealing with people or communities that are already marginalised or stigmatised in some way.

STORY BOX: 'Thinking together' in a garbage dump

Natalia Luxardo is a social worker who spent five years doing community-based research with group of waste pickers and their families, in an open-air garbage dump in Argentina. Many visual stereotypes of this community already existed, including from several CNN and BBC reports about the site. This meant that when it came to sharing their findings with the world, the research team had some difficult decisions to make, particularly around their use of film. They felt they had a responsibility not to re-embed structural violence through repeating negative and stereotyped images. But if they tried to avoid emphasising the miseries, there was the risk of presenting a "falsely romantic" view which would go against their responsibility "to 'witness' the infringement of basic rights" that the conditions they had documented represented (Luxardo 2022, p. 7). Since the project was embedded in an action research approach, the researchers and participants practiced "thinking together" through workshops in which they tried to decide what should be included in or removed from the documentary. These workshops doubled as part of an ongoing consent process that dealt with issues of confidentiality and anonymisation. They settled on something that could include "confronting" images and harsh conditions, but also include "the waste-pickers' daily resistances, expectations, dreams, social networks, and so on": aiming then to represent "the whole picture" (9). Because they were worried about not being able to anticipate or control all the channels through which the film might be disseminated, they settled on not having it available publicly, focusing instead on settings where it could be presented to audiences "within a clear context and framework", for example, as part of a seminar or conference presentation.

Ethnographers have often been drawn towards studying the difficult aspects of human existence, including people living in extreme situations, or who have encountered significant challenges (Robbins 2013). In these cases in particular, care is needed to avoid entrenching otherness, or a sense of the abject, through an obsession with suffering. *How can a sense of agency, dignity, humanity, and individuality be retained when representing participants who are also subject to personal suffering or structural violence?*

Texts that lay out things that are explicit or horrific must consider their responsibility to audiences as well as participants. *What is the intended impact on audiences? What might the unintended impact be?* There are similar responsibilities when the ethnographer is speaking in a personal or autoethnographic mode about their own painful experiences. This type of work, Behar argues, must also be purposeful – not just exposure for its own sake – and must reflect on the responsibility of the researcher to not traumatise or retraumatise audiences, and to care for both others and themselves (Behar 1996). This dovetails with arguments for a turn towards research that focuses on care, empathy, hope, and the many ways in which people are "working to construct a liveable world" (Robbins 2013) even amidst enormous challenges.

SAFETY, ANONYMITY, AND IDENTIFICATION

An ethnographic text does not only impact abstract audiences 'out there'. It can also affect the individuals that are represented in it (i.e. the research participants) when they read or view it themselves, or when others in their social circle do. "We make personal choices about what to include and omit from our accounts: whether to include gossip, confidential and sensitive information, the dark side of organizational life, potentially damaging information to participants or the organization" (Cunliffe and Alcadipani 2016, p. 554). These choices must be made with consideration to participants' physical safety, financial prospects, social status, and so on.

STORY BOX: Return to the drag queens

Historian Leila Rupp and sociologist Verta Taylor, both working feminist studies, conducted ethnographic research with drag queens in Florida in the late 1990s. Two years after they finished their fieldwork, they published a book (Rupp and Taylor 2015). Then, nearly a decade after the original project finished, they went back again to their fieldsite. The original participants remained proud of the book, even referring to it as 'our book'. But Rupp and Taylor were also able listen to participants' stories about the impact the book's publication had made on their lives, including details about what they wished had or hadn't been in there, and what friends and family had thought about it. This involved bigger consequences, such as when family members were upset – and in fact "screaming" as one person put it – from finding out they were drag queens. Rupp and Taylor also checked in about issues surrounding their mention of drug use in the book – a difficult issue, given the small-town setting and the chance of it bringing illegal activities to the notice of the police (Rupp and Taylor 2011). More generally, their participants reflected on the way that their lives and they themselves had evolved and changed since the book's publication, some of them wishing they could "add an addendum" to the stories it told.

Many of the practices emphasised by institutional ethics committees – such as offering participants the opportunity to be anonymised – are meant to mitigate possible harms from this process. There is a particular challenge in this for ethnographers. It is one thing to anonymise someone by giving them a pseudonym, but this is not the same as de-identifying them totally. To put a participant's words or stories in context, an ethnographer may often want to include at least *some* information about the person's background, positionality, and wider life. Ethnographers will, ideally, engage with participants through multiple layers of reflexive conversation to establish what level of identifiability feels comfortable and safe to them. The practice of creating 'composite characters' – which involves using an aggregate of multiple people's stories, experiences, or data to represent broader trends or findings – can help with storytelling that doesn't rely on identifying actual individuals. However, seeing the details of their own experiences mixed in with those of others might also be disconcerting for some people.

Indeed, some participants might *want* to be named and identified as an expression of agency. Australian anthropologist Sophie Chao shared how this came up in conversation with her participant, Marcus, who responded emphatically to the idea of anonymisation by saying, "The government and corporations have taken our land and forests [. . .] They have taken our food and future. We have lost everything. Yet still, you would take away our names?" (Chao 2022, p. 27). My own stance is that people have a right to claim their own stories in their own names (with fully informed consent about when and how these are being used), which should never be overridden by institutional policies that prefer anonymisation as a blanket rule for sensitive topics.

> **STORY BOX: The public face of health policy**
>
> CGMs (continuous glucose monitors) can be a life-saving device for people with type 1 diabetes, but they were not publicly funded in Aotearoa New Zealand until mid-2024. This followed a long period of public activism and advocacy. In 2021, wanting to contribute to this, I committed to some work on this topic as part of a larger research grant focused on medical crowdfunding. Based on my activist stance, as soon as I had some findings, my priority was to work out how to communicate them to a public audience. I settled on a long-form essay on a popular news and culture website. But I knew this site was successful largely because of its eye-catching graphics. What were my options, given that it seemed like the best way to humanise the issue was to put a face to it, but I did not want my participants to be directly identifiable? As soon as I had the idea to produce digital illustrations, I discussed it with the families from my case studies, who were open to the suggestion and provided some additional photographs for me to work from. When I shared drafts of both the artwork and the article with them, the positive responses and the requests to use their actual first names (rather than the pseudonyms that I had suggested) reassured me that they felt okay with the way I was representing their stories in a public space.

COMMUNICATING STRATEGICALLY, WIDELY, AND ACCESSIBLY

Most ethnographers want their work to be useful. Many have particular goals or ideas around directions of social or structural change that would benefit their participants, or the wider world. It can be beneficial

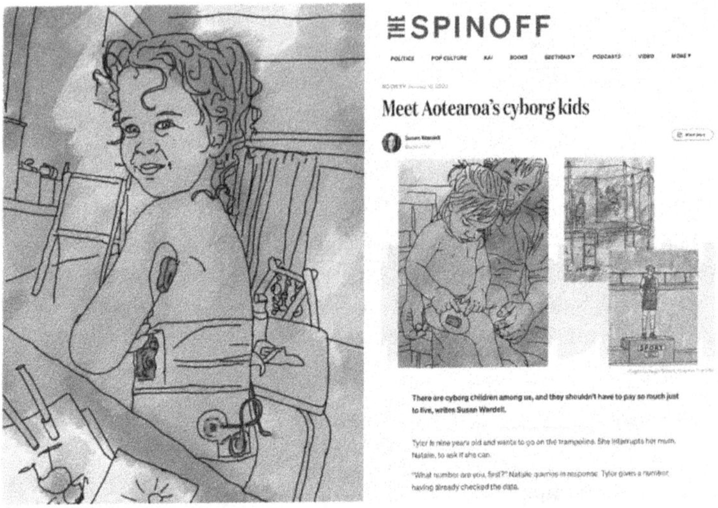

Figure 8.6 Left: example of digital artwork produced by drawing over photos of participants and adding light colourisation. Right: the drawings as presented on popular news and culture website *The Spinoff* in January 2022.

to think actively and from early stages about how your findings can be made accessible and compelling to policy-makers, community leaders, or other major stakeholders. One challenge can be that many of these people have their own 'language', or conventions of communication, which do not necessarily fit well with the nuanced, long-form type of text ethnographers typically produce. Some stakeholders might be used to graphs or statistics, technical reports, or two-minute briefings, for example. Ethnographers working on multidisciplinary teams or working directly with policy-makers may develop skills in 'translating' their work for different audiences, or settings. They need to work out how to continue to honour the ethnographic lens, and the nuance and complexity of their data, when they do.

Policies, behaviours, and beliefs, can also change over time through public engagement. To educator and feminist scholar bell hooks, writing complex concepts in more accessible ways was a radical and political act. In some ways, ethnographic texts are uniquely placed to be engaging and accessible for audiences well beyond an academic context because they are humanistic as well as scientific.

They also deal with many things that matter imminently to a lot of people. For all of these reasons, ethnographers can (and do) do the work of public engagement. This can take a variety of forms. It may mean writing compelling public-facing books, or writing for the media and/or for digital platforms. It may mean drawing on creative, visual, or multimodal forms that are accessible and interesting for the public too, as I have described earlier in the chapter.

Key challenges can come not only from deciding on the content or style of ethnographic texts but from grappling with their systems of production. Academic publishing and the peer-review process typically make for a long timeline between producing a text and publishing it. It may also put up barriers to access such as high purchase cost or paywalls, with the rise of more open-access forms of publishing only a partial (though important) solution to this. 'Quicker' mediums, such as blogging and other forms of online publishing, may allow authors to provide a more timely contribution to public conversation. However the ability to speak meaningfully into public debates as they happen can sit in tension with the rigor and care required to attend to our responsibilities in representing real people and complex situations, simply because it takes time to do this well (Bonilla 2020). Speaking out on certain topics may have unpredictable political or professional consequences for the ethnographer too.

STORY BOX: Screening women's songs

As an upper-caste man from New Delhi, Rajat Nayyar hadn't been aware of the diversity of vocal traditions in the performances he was trying to document when he began his project in rural North India. Nor was he fully aware of the unequal representations of women from various caste groups. In 2015, his ethnographic project brought him into a collaborative relationship with Aaji, a lower-caste storyteller, singer, healer, and rice farmer. Over the next three years, as Nayyar worked on a multimodal project about singing traditions and political activism, he and Aaji collaborated to put recordings of her songs on YouTube. She regularly asked for updates on the number of views and comments that these received. After editing together his own ethnographic videos and films, he returned to the village in 2018 to screen these to the community. Aaji herself hosted one of these events and sat beside the laptop as it played footage of her singing to an audience of people from a variety of different castes (Nayyar 2022).

Figure 8.7 Aaji presents the documentary film to the Badhuli community, Bihar, 2017.

Source: Photo credit: Rajat Nayyar.

People have a right to access their own data and stories in whatever form we turn them into. Communicating findings back to participants in an accessible way, and ideally in a way that brings actual benefits to the community, is another key ethical responsibility for the ethnographer. Depending on the field, simply handing participants a copy of our reports or academic writing outputs may not actually achieve this. There are a range of other options, however. Daniel Miller has recently argued for an 'ethnography of dissemination', suggesting that decisions about how to share and distribute findings should come from using our existing strengths to establish a close *ethnographic* attention to local settings and audiences. This can include thinking about how different genres might take on authority in different ways in these contexts, and what people normally enjoy or value in these settings. It is helped by thinking about the affordances of certain platforms, mediums, or genres for meeting not only the ethnographer's goals, but the goals that participants and collaborators have in their own lives and communities. While this is a lot to factor in, it all works towards ethnographic texts that can truly contribute to a tradition of engaged, ethical, inclusive, and impactful work.

CHAPTER SUMMARY: KEY POINTS

- Crafting texts is an important part of an ethnographer's work, and involves many layers of thinking and sense making. This can be a complex, time-consuming task, but also a creative and joyful one, and many ethnographers identify as storytellers.
- Ethnographers work in a variety of mediums and genres based on what is fitting for their topic and their audience. Each has different affordances for bringing to life vivid, nuanced representations of a particular place, time, and people. Each also has different practical, ethical, and epistemological challenges.
- Practices of ethnographic representation have shifted over time and in response to debates about objectivity, authority, truth, and voice. Ethnographers must grapple with the process of speaking about, for, or *with* other people. There are a variety of ways ethnographers can co-produce texts, with participants.
- Ethnographers should strategically consider their responsibility for how ethnographic knowledge is used and circulated in the world, including its impact on participants, policy-makers, and the public.
- Ethnographic texts can make a meaningful impact on the world, cultivating curiosity and empathy, challenging norms, asking people to see things differently, and providing a platform for more voices and perspectives.

RECOMMENDED FOR FURTHER READING

Alfonso, A.I., Kurti, L. and Pink, S. (2004). *Working images: visual research and representation in ethnography.* London: Taylor & Francis Group.

Illustrating Anthropology Website. https://illustratinganthropologycom.wordpress.com/.

McGranahan, C. (2020). *Writing anthropology: essays on craft and commitment.* Durham: Duke University Press. Available from: https://doi.org/10.2307/j.ctv123x7rf.

Shah, A. (2022). Why I write? In a climate against intellectual dissidence. *Current Anthropology*, 63(5), pp. 570–600. Available from: https://doi.org/10.1086/722030.

Stinnett, J. (2012). Resituating expertise: an activity theory perspective on representation in critical ethnography. *College English*, 75, pp. 129–149.

REFERENCES

Abu-Lughod, L. (2000). Locating ethnography. *Ethnography*, 1(2), 261–267.

Alfonso, A.I., Kurti, L. and Pink, S. (2004). *Working images: visual research and representation in ethnography*. London: Taylor & Francis Group. Available from: http://ebookcentral.proquest.com/lib/otago/detail.action?docID= 200271 [accessed 8 October 2023].

Behar, R. (1996). *The vulnerable observer: anthropology that breaks your heart*. Boston: Beacon Press.

Behar, R. (2020). Read more, write less. In: McGranahan, C., ed. *Writing anthropology*. Durham: Duke University Press (Essays on craft and commitment), pp. 47–53. Available from: https://doi.org/10.2307/j.ctv123x7rf.10.

BOMB Magazine. (no date). Rehearsing violence: Nomi Stone interviewed. *BOMB Magazine*. Available from: https://bombmagazine.org/articles/nomi-stone-kill-class/ [accessed 30 January 2024].

Bonilla, Y. (2020). Quick, quick, slow: ethnography in the digital age. In: McGranahan, C., ed. *Writing anthropology*. Durham: Duke University Press (Essays on craft and commitment), pp. 118–120. Available from: https://doi.org/10.2307/j.ctv123x7rf.23.

Chao, S. (2022). *In the shadow of the palms: more-than-human becomings in West Papua*. Durham: Duke University Press. Available from: https://doi.org/10.2307/j.ctv2j86bm4.

Chio, J. (2021). Visual anthropology. In: *Cambridge encyclopedia of anthropology* [Preprint]. Available from: https://www.anthroencyclopedia.com/entry/visual-anthropology [accessed 12 June 2023].

Cunliffe, A.L. and Alcadipani, R. (2016). The politics of access in fieldwork: immersion, backstage dramas, and deception. *Organizational Research Methods*, 19(4), 535–561. https://doi.org/10.1177/1094428116639134.

Daly, K. (1997). Re-placing theory in ethnography: a postmodern view. *Qualitative Inquiry*, 3(3), pp. 343–365. Available from: https://doi.org/10.1177/107780049700300306.

Dattatreyan, E.G. (2015). Waiting subjects: social media–inspired self-portraits as gallery exhibition in Delhi, India. *Visual Anthropology Review*, 31(2), pp. 134–146. Available from: https://doi.org/10.1111/var.1207.

Deger, J., Saxena, A.K. and Zhou, F. (2021). Introduction to feral atlas. In: Tsing, A., Deger, J., Saxena, A.K. and Zhou, F., eds. *Feral atlas: the more-than-human Anthropocene*. Redwood City: Stanford University Press.

Falcone, J.M. (2020). Genre bending, or the love of ethnographic fiction. In: McGranahan, C., ed. *Writing anthropology*. Durham: Duke University Press (Essays on craft and commitment), pp. 212–219. Available from: https://doi.org/10.2307/j.ctv123x7rf.44.

Geertz, C. and Darnton, R. (1973). Thick description: toward an interpretive theory of culture. In: *The interpretation of cultures*. New York: Basic Books.

Gilbert, A. and Kurtović, L. (2022). Labours of representation: a Bosnian workers' movement and the possibilities of collaborative graphic ethnography. *Anthropologica*, 64(1), pp. 1–34. Available from: https://doi.org/10.18357/anthropologica6412022361.

Hage, G. (2020). Antiracist writing. In: McGranahan, C., ed. *Writing anthropology*. Durham: Duke University Press (Essays on craft and commitment), pp. 149–152. Available from: https://doi.org/10.2307/j.ctv123x7rf.31.

Hartblay, C. (2024). *I was never alone*. Available from: https://cassandrahartblay.net/i-was-never-alone/.

Ingold, T. (2011). *Redrawing anthropology: materials, movements, lines*. Farnham: Ashgate Publishing, Ltd.

Luxardo, N. (2022). Ethics in practice and ethnography: faux pas during fieldwork with structurally vulnerable groups. *Medicine Anthropology Theory*, 9(3), pp. 1–13. Available from: https://doi.org/10.17157/mat.9.3.5747.

McGranahan, C. (2015). What is ethnography? Teaching ethnographic sensibilities without fieldwork. *Teaching Anthropology*, 4. Available from: https://doi.org/10.22582/ta.v4i1.421.

McGranahan, C. (2020). Writing about bad, sad, hard things. In: McGranahan, C., ed. *Writing anthropology*. Durham: Duke University Press (Essays on craft and commitment), pp. 131–133. Available from: https://doi.org/10.2307/j.ctv123x7rf.26.

McGranahan, C. (2022). Theory as ethics. *American Ethnologist*, 49(3), pp. 289–301. Available from: https://doi.org/10.1111/amet.13087.

Meyer, M. (2017). Re-framing the ethnographic encounter: *Les Maîtres Fous. Journal Des Africanistes*, (87–1/2), pp. 198–220. Available from: https://doi.org/10.4000/africanistes.5553.

Mol, A. (2003). *The body multiple: ontology in medical practice*. Durham: Duke University Press (Science and Cultural Theory).

Murphy, E. and Dingwall, R. (2001). The ethics of ethnography. In: *Handbook of ethnography*. London: SAGE Publications. Available from: https://doi.org/10.4135/9781848608337.

Nayyar, R. (2022). Granular activisms. *Visual Anthropology Review*, 38(2), pp. 256–278. Available from: https://doi.org/10.1111/var.12277.

O'Reilly, K. (2012). *Ethnographic methods*. Florence: Taylor & Francis Group. Available from: http://ebookcentral.proquest.com/lib/otago/detail.action?docID=958470 [accessed 28 September 2022].

Pandian, A. (2019). *A possible anthropology: methods for uneasy times*. Durham: Duke University Press. Available from: https://doi.org/10.2307/j.ctv11sn173.

Robbins, J. (2013). Beyond the suffering subject: toward an anthropology of the good. *Journal of the Royal Anthropological Institute*, 19(3), pp. 447–462. Available from: https://doi.org/10.1111/1467-9655.12044.

Rupp, L.J. and Taylor, V. (2011). Going back and giving back: the ethics of staying in the field. *Qualitative Sociology*, 34(3), pp. 483–496. Available from: https://doi.org/10.1007/s11133-011-9200-6.

Rupp, L.J. and Taylor, V. (2015). *Drag queens at the 801 cabaret*. Chicago: University of Chicago Press. Available from: https://press.uchicago.edu/ucp/books/book/chicago/D/bo3643388.html [accessed 22 February 2024].

Shah, A. (2022). Why I write? In a climate against intellectual dissidence. *Current Anthropology*, 63(5), pp. 570–600. Available from: https://doi.org/10.1086/722030.

Stinnett, J. (2012). Resituating expertise: an activity theory perspective on representation in critical ethnography. *College English*, 75(2), pp. 129–149.

Stone, N. (2020). Poetry and anthropology. In: McGranahan, C., ed. *Writing anthropology*. Durham: Duke University Press (Essays on craft and commitment), pp. 195–200. Available from: https://doi.org/10.2307/j.ctv123x7rf.41.

Van Maanen, J. (2006). Ethnography then and now. *Qualitative Research in Organizations and Management: An International Journal*, 1(1), pp. 13–21. Available from: https://doi.org/10.1108/17465640610666615.

Van Maanen, J. (2010). A song for my supper: more tales of the field. *Organizational Research Methods*, 13(2), pp. 240–255. Available from: https://doi.org/10.1177/1094428109343968.

Watson, A. (2022). Writing sociological fiction. *Qualitative Research*, 22(3), pp. 337–352. Available from: https://doi.org/10.1177/1468794120985677.

White, N. (2010). Indigenous Australian women's leadership: stayin' strong against the post-colonial tide. *International Journal of Leadership in Education*, 13(1), pp. 7–25. Available from: https://doi.org/10.1080/13603120903242907.

Wolcott, H.F. (2008). *Ethnography: a way of seeing*. California: AltaMira Press. Available from: http://ebookcentral.proquest.com/lib/otago/detail.action?docID=1343764 [accessed 28 September 2022].

EPILOGUE
A love letter to ethnography

Dear readers, dear ethnographers,

Throughout its 100-plus-year history, ethnography has gone on many journeys. It has wrestled, writhed, transformed itself, and spread into a variety of different settings, to continue strong as a unique mode of inquiry into social and cultural life. By way of summarising some of what I have presented about this in the previous chapters, and concluding on a more personal note, this short chapter stands as my love letter to ethnography. While for most of the book, I have not often drawn explicitly on my own experiences, here I speak based entirely on these, while acknowledging of course that my experiences are not representative of everyone's.

I focus on what ethnography brings to the table. I speak less about specific techniques and more about the broad values I see as driving ethnographic practice. This discussion is organised into six points. These, I admit, present an idealised vision of ethnography while recognising, as I hope I have done throughout, that ethnographers don't always meet their own ideals, and it remains a methodology replete with its own tensions and paradoxes. You can think of what follows, then, as a 'best-case scenario': an act of celebration, rather than a persuasive act, and a map of hope and possibility for the ethnographers of the future.

1 Engaging curiously in a plural world

Our world is full of others. Our world is full of *worlds*. It is 'incorrigibly plural', as one poet said,[1] and that is both exhilarating and terrifying. This encompasses diverse and overlapping cultures, subcultures, identities, genders and sexualities, family structures, institutions, professions, beliefs, political alignments, and more. Amidst enormous complexity, and in an age of many sorts of tension and crisis, we need to know how to live, work, play, love, govern, educate, and thrive with and alongside others of every sort. Ethnography *began* with an interest in others, and it continues to work actively to deconstruct some of the colonial formulations of power that shaped its approach to this. But today, it remains more situated than ever in the radical possibility of challenging any singular view of the world and embracing, instead, multiplicity and plurality. As such, it feeds our (understandable, allowable) curiosity about those around us but, with efforts to avoid sensationalism, fetishisation, or other*ing*.

As an ethnographer, I have constantly been astonished by and grateful for the opportunities to engage with those very different from myself – in interviews, in participant-observation, even just in reading ethnographic texts. I have felt myself move through discomfort, disengagement, or suspicion to become enriched and expanded. Ethnography has never insisted I set aside my own (moral, political, and religious) commitments or identities to do this. I *have* been called towards deep reflexive work in order to become aware of the assumptions and privileges of my own worlds; to practice defamiliarisation, and analyse myself as deeply as I have my participants. But though this has been challenging, it has not been *harmful* or excluding. Rather, I have been able to come as myself and invite others to do the same through practices of observing, listening, and documenting that took differences (including my own) as interesting and valuable rather than threatening.

2 Something you can put weight on

Ideas are powerful. Stories are powerful. The representations and knowledge we have about each other can shape and reshape the world. Ethnography remains committed to providing an *empirical* basis for discussions about culture and society, diversity and difference, power and the everyday. This certainly doesn't mean that ethnographic data is infallible or that it can't use creative or speculative techniques too. But it is also based on more than ideas that sound pretty or compelling or convenient at the time. Instead, it is connected to transparently conducted practices of observation, recording, analysis that are open others to examine and critique. I have been grateful to be able to hand a technical report to a diabetes patient advocacy organization, to write op-eds about the memorialisation of a white supremacist attack for a national news website, and to present a workshop on social media admins to a government group, with this type of rigor behind me.

The measures of validity, legitimacy, and rigor in ethnographic research can be different from those of many other empirical fields, being based on reflexivity, transparency, and critical thinking, rather than the objectivity demanded of positivistic work. It is a form of rigor nonetheless. It means the knowledge we produce can bear weight, can be put to work, and at the very least is accountable for what it contributes to the world.

3 Counter-cultural paths to knowing

Knowledge-making is itself an industry. It is often subject to pressures to *produce*, in certain time frames and in certain (tidy, authoritative, and auditable) formats. As I have worked my way through academia, I have found this both tiring and tiresome, and I don't doubt that other commercial and applied research settings have similar if not greater pressures. But many of the core principles of ethnographic research run against this. Depth over breadth. Rich data over quantities of data. A long-term and committed approach to learning *with* rather than just *about* people. In institutional settings, where it is often difficult to justify extra funding or time, 'ethnography' is a magic word to explain why you need to take the long path, and we have the weight of all these years of tradition and debate to reassure ourselves and others that the proof is in the pudding. It works.

It's a paradox, really, that this *smaller* focus (on a very particular place, time, group, or practice) actually allows for a *bigger* lens, acknowledging the interconnectedness of . . . well, everything. While ethnographers today reject the idea that they can describe their fields in total, they also refuse to carve the social world up into tiny pieces or become too specialised, saying, *Language is connected to politics. Kinship is connected to economics. Work is connected to religion. Laundry is connected to . . . everything.* Ethnography asks for time to unpack this with the people most closely involved. And it refuses, even with all of this time, to make sweeping or simplified claims about what it has found, instead providing complex, nuanced descriptions of how the various parts of the social world interact in the personal lives of its participants and insisting that these, too, *matter.*

4 Flexible techniques for complex lives

The contexts in which researchers are expected to enact their research keep changing, from new funding models to restructured institutions to global pandemics. At the same time, researchers themselves are living human lives: having families, experiencing health and illness, facing precarious financial situations. Amidst and against all this, ethnographic research is not easy to do. It is time-consuming, demanding, and unpredictable. It is driven by very high ethical standards. But at the same time, ethnographers can be flexible about the way they set up their studies and the techniques they use. This means they can adapt to work inclusively with different people and communities in a variety of social contexts. It also means that more people, with diverse situations and dispositions, can *become* ethnographers. Indeed some of the same things that make ethnography challenging do also create a sort of grace for the ethnographer themselves. It's an open-ended, flexible, and iterative process that provides a space to adapt to their own fumbles, to learn and return, to change course if needed. It enables you to chase down the wonderful thing that presented itself, the interesting thing you *could* learn, rather than being stuck beating your head against the wall of what you *thought* you would learn or the way you thought you would learn it. It can make research feel like a pool of infinite possibility, offering permission to innovate and adapt if the task calls for it and to follow the things that spark and sing. It can also make research feel kinder, and more possible. In my own career, this flexibility has allowed me to find ways to continue my work within the parameters of being an adjunct with little to no funding, being a mother of small children, being a full-time academic with teaching responsibilities, being neurodivergent, being chronically ill, and many other challenging life circumstances, as well as helping me respond to unforeseen twists in the research process.

5 Relational affects and relational effects

Ethnographers dare to tread where few are brave enough to go, and where the personal, professional, and social stakes are highest. This can include researching extremely sensitive, emotional, fraught, politicised, and complex facets of human social and cultural life. But we do this in a way in which we are face-to-face with our participants (and sometimes in which we are one of them). In these scenarios there is nowhere to hide, but rather we must account for what impact both our practices and our outputs have in the world. This requires care, commitment, and bravery. Ethnographers have to go beyond ethical box-ticking to do this and, instead, be willing to form relationships with individuals and communities. It takes a lot of extra time, effort, and energy, which, again, can run counter to rushed institutional processes or timelines. At times, I have felt myself chaffing at the extra layers of commitment and responsibility that come from forming *actual* relationships in the field, with obligations that extend long after. But I also value the fact that I don't get to just walk away. It is a reminder that what we do is real, and it matters to real people in real worlds.

These relationships also increase the likelihood that the work will shape the ethnographer themselves. It allows and acknowledges research to be affecting. I have been humbled by being handed food by an earnest youth worker in small concrete building in Kampala city. I have felt guilt creeping up my fingertips from the screen as I scrolled through a teenage girl's crowdfunding page. I have cried with a fellow mother as she recounted the day of her child's near-fatal shooting, just a few blocks from where we sat. We learn to take as meaningful and valid, the experiences and wisdom of others. We learn to carry the stories of those we meet and, as such, become entangled with their pasts and futures as we work out how to honour and care for those stories in their afterlife as ethnographic knowledge.

6 The messy, beautiful whole

It really is a gorgeous, big, juicy mess, this human existence. Especially when we consider not only the longings and tremblings of the individual human person but the vivid, aching jumble of our *shared* life in communities, organisations, cultures, and societies. Many different people in many different fields are drawn to study this. Sometimes, for practical reasons, they have to squeeze and distil knowledge about the social world into forms that don't much resemble their origin: for example into statistics, graphs, diagrams, and models. These serve their own purposes, of course. But I celebrate ethnography as a practice which works towards analysing the social world without sucking the life out of it.

While remaining grounded in empirical data, ethnographers deal in textured, colourful description, in stories, in forms that aim to retain a sort of wholeness. This, in turn, leaves room for mess, nuance, and complexity. For incompleteness, partiality. For tension, paradox, and contradiction. In other words, for 'humanness'. This includes making room for the ethnographer, too, to be more than a 'brain on a stick'. In working to represent the types of worlds that you can touch, feel, taste, and perhaps even recognise – in allowing ourselves to be storytellers, communicators, artists, and whole people – ethnographers position themselves well to be able to communicate the insights they gather to audiences well beyond the ivory tower. As such, they can engage wider publics in expanding their understandings of humanness at its best and its worst, and in all it's plural forms. Which brings us just about full circle.

So there you have it: six sonnets in a love letter to ethnography, along with many more reasons for appreciating this (challenging, imperfect, still-evolving) research methodology sprinkled throughout the book.

Thank you for listening, ethnographers, others, and all. I wish you well in all of your ethnographic endeavours.

With aroha
Susan

NOTE

1   Louis MacNiece. (1967). Snow. Available from: https://www.poetryfoundation.org/poems/91395/snow-582b58513ffae.

INDEX

For Product Safety Concerns and Information please contact our EU
representative GPSR@taylorandfrancis.com
Taylor & Francis Verlag GmbH, Kaufingerstraße 24, 80331 München, Germany

www.ingramcontent.com/pod-product-compliance
Lightning Source LLC
Chambersburg PA
CBHW050646270326
41927CB00012B/2895

* 9 7 8 1 0 3 2 5 2 0 1 2 4 *